Stanley Wood

Over the Range to the Golden Gate

Stanley Wood

Over the Range to the Golden Gate

ISBN/EAN: 9783337400095

Printed in Europe, USA, Canada, Australia, Japan

Cover: Foto ©Andreas Hilbeck / pixelio.de

More available books at **www.hansebooks.com**

OVER THE RANGE

TO

THE GOLDEN GATE.

A COMPLETE TOURIST'S GUIDE

TO

COLORADO, NEW MEXICO, UTAH, NEVADA, CALIFORNIA,
OREGON, PUGET SOUND AND THE
GREAT NORTH-WEST.

BY STANLEY WOOD.

CHICAGO:
R. R. DONNELLEY & SONS, PUBLISHERS
1891

PREFACE.

T is no light undertaking to prepare a guide book which shall adequately describe the places of interest on the great trunk lines between Denver on the hither side of the Rocky Mountains, San Diego at the southern extremity of California, and Portland, Seattle and Tacoma, the three commercial entrepots of the Great Northwest. Yet such is the undertaking purposed: In a work of this character fact must ever stand paramount to fancy, and lucidity of expression take the precedence. No attempt will be made at "fine writing;" every effort will be made to state just such facts as the traveler would like to know, and to state these facts in clear and explicit language.

The country traversed is most interesting, abounding in scenes of the greatest variety, from the broad and billowy expanses of the boundless prairie to the rugged grandeur of the American Alps, from the picturesque quaintness of New Mexico and the nomadic wildness of the Indian reservations to the polished civilization of metropolitan cities. There is no journey which can be taken on the continent of North America that presents so much of interest to the tourist, and which can be taken with such a comparatively moderate outlay of time and money, as the one described in the following pages. New Mexico, Colorado, Nevada, California, Oregon, Washington Territory! What a field for investigation, investment or pleasure! These are the lands of gold, of silver, of coal, of agriculture, of all fruits known to the temperate and sub-tropical zones These are the lands of new endeavors, of fresh impulses, and for these reasons are of special interest to tourists business men and seekers after health and pleasure. Aside from the interesting character of the subject discussed, there is also a special value in the work now presented to the reader, inasmuch as great care has been taken to gather information that shall be found statistically accurate. In a work of this character it is difficult to combine accurate information with matters of general interest in such a way that neither shall have an undue prominence. The writer has endeavored to attain this desirable medium. One thing is certain, nothing in this book is venal in its character. The opinions here expressed are those of the writer; the descriptions of scenes given here are reproductions of the feelings inspired by those scenes. There has been no bias in any direction. On the contrary, every effort has been made to write judicially and, at the same time, retain the enthusiasm which the traveler naturally feels in beholding new sights and scores.

In order that no element of information may be lacking, carefully prepared tables of statistics have been given a place in this volume, and the reader is respectfully requested to make use of these tables because much of value has been condensed into this convenient form

By the aid of the tables referred to, and by frequent reference to the three excellent maps herein given, the tourist will be able to gain an exceptionally

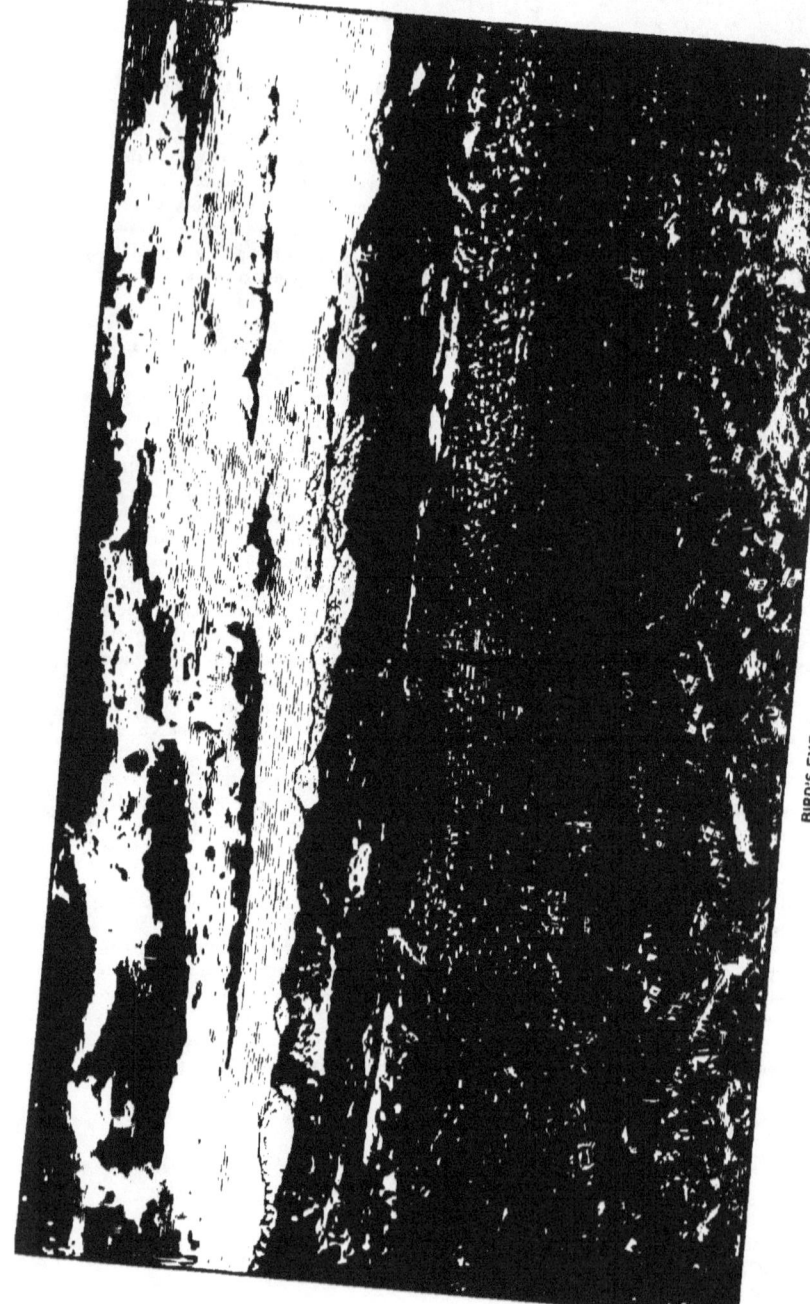

BIRD'S-EYE VIEW OF DENVER.

clear idea of the journey he is making, and of the characteristics of the country through which he is passing.

As another aid to the traveler abundant illustrations have been prepared, which will give the purchaser of this book an idea of what he may expect to see; and which, after he has beheld these places, will serve as a reminder of those pleasant scenes which by their assistance can never fade from his memory.

It has been the endeavor of the writer to meet as nearly as possible the wants of all classes of travelers. Information of value to the tourist for pleasure, the health seeker, the sportsman and the man of business, will be found in the pages of this book. Nothing has been written in the interests of any clique or class. The truth, and nothing but the truth, has been told. If there are errors they are such as must necessarily occur in the compilation of a work covering such a vast extent of territory. Accuracy has been aimed at, and as a whole, the writer can vouch for the accuracy of what will be found herein. The book is one written in the field and not in the study. Facts are not taken at second hand. The author writes of what he saw with his own eyes, and not what he read. The statistics have been gathered from authentic sources, and have been condensed into the most compact and convenient form. Hoping the book may prove a useful companion to the traveler, it is submitted without further comment to the public.

VIEW OF FOURTEENTH STREET, DENVER.

FROM THE MISSOURI RIVER TO DENVER.

THE Missouri River has come to be regarded in a general way, as the boundary line between the East and the West, although, in truth, the terms east and west are extremely elastic in their application. However, for the purposes of this book we will consider that all on the sundown side of the Missouri River is West, and that the traveler has reached one of the three great entrepôts to this vast country and finds himself in Omaha, St. Joseph or Kansas City. From either of these thriving cities the journey to Denver can be taken by way of first class transportation lines provided with all the modern conveniences and luxuries.

From Omaha one has choice of the Burlington route and the Union Pacific, and from Kansas City one can travel by either of the above lines with an additional choice between the Missouri Pacific, the Rock Island, or the Atchison, Topeka & Santa Fe railroads. With Chicago or St. Louis as the initial point one can go direct by any of the trunk lines to the Missouri River and continue his journey to Denver over his choice among the routes mentioned above.

The trip across the great plains from the Missouri River to Denver is full of interest and variety to one who beholds this vast expanse for the first time. Nothing can give such a vivid impression of the greatness of our country, and the adventurous character of our people, as the sight of these boundless prairies and the habitations of the hardy pioneers who are rapidly turning the buffalo sod and exposing the rich black soil to the fertilizing action of the sun and air, and substituting for nature's scant forage, abundant harvests of corn and wheat. The railroads for a distance of three or four hundred miles to the west of the Missouri River, pass through thriving cities to which a comparatively thickly settled agricultural country is tributary. Then the newer territory is reached, the towns are of less frequent occurrence and smaller in size, the plains appear more nearly in their native state, only dotted here and there with the claim cabins of the settlers. As the traveller looks out of the car window across the billowy expanse, he sees herds of cattle and sheep, grazing on the rich bunches of buffalo grass, and occasionally he will catch a glimpse of the flying form of an antelope disappearing over the brow of a distant rise of land. Not uninteresting are the prairie dog villages with their preternaturally grave inhabitants sitting on their haunches like diminutive kangaroos, and the writer has seen a whole car load of people filled with the most pleasurable excitement over the efforts of a jack rabbit to outspeed the iron horse. With these and many other novel and interesting sights the time is whiled away

THE HIGH SCHOOL BUILDING, DENVER.

until some traveler, more experienced, or more sharp of sight, suddenly cries out "The Mountains!" There is a rush to his side of the car and everybody gazes earnestly, and amidst eager exclamations and doubting comments the blue of the sky is at last disintegrated from the blue of the mountains, and the most skeptical at length acknowledges that the stain of ultramarine, with its undulating sweep against the western horizon is really the distance-enchanted range of the Rocky Mountains. Soon patches of fleecy white appear, and with a sigh of disappointment the traveler decides that the clouds are dropping down and will soon shut out the view of those "sentinels of enchanted land," but gazing more intently, it dawns upon the mind at last that those glimmering expanses are not veils of cloud, but are in fact mountain fields of everlasting snow! The Snowy Range has at last declared itself, and from this moment until the trans-continental journey shall have been accomplished, the traveler will have the immediate memory or the intimate presence of the mountains with him continually.

The view of the Rocky Mountains, which the traveler gains on approaching Denver from the east is one of unsurpassed beauty, and that this statement may not rest on the dictum of this book, let us take the testimony of the greatest traveler, and the most graceful descriptive writer America has yet produced. Bayard Taylor says :—"I know no external picture of the Alps that can be placed beside it. If you take away the valley of the Rhone, and unite the Alps of Savoy with the Bernese Overland, you might obtain a tolerable idea of this view of the Rocky Mountains. Pike's Peak would then represent the Jungfrau, a nameless snowy giant in front of you, Monta Rosa and Long's Peak, Mount Blanc. The altitudes very nearly correspond, and there is a certain similarity in forms. The average height of the Rocky Mountains, however, surpasses that of the Alps. * * * From this point there appears to be three tolerably distinct ranges. The first rises from two to three thousand feet above the level of the plains, is cloven asunder by the cañons of the streams, streaked with the dark lines of the pine, which feather its summits and with sunny, steep slopes of pasture. Some distance behind it appears a second range, of nearly double the height, more irregular in its masses, and of a dark velvety violet hue. Beyond, leaning against the sky, are the snowy peaks, all of which are from thirteen to (nearly) fifteen thousand feet above the sea. These three chains, with their varying but never discordant undulations, are as inspiring to the imagination as they are enchanting to the eye. They hint of concealed grandeurs in all the glens and parks among them, and yet hold you back with a doubt, whether they can be more beautiful near at hand than when beheld at this distance."

The doubt so gravely expressed in the last sentence of our quotation, the traveler, when he shall have taken the trans-continental tour, will be fully able to resolve for himself. He will have beheld a bewildering variety of beauty, and in the quiet evenings at home, he will find material for the most exquisite enjoyment of pleasing reminiscence and reverie.

With such an approach, Denver must needs be something more than ordinary not to strike the traveler as a discord in the grand harmony of the scene. It is a fact, and it is a pleasure for the writer to record it, that Denver is never a disappointment. What its peculiar charms may be, and how it appears to the stranger within its gates, will be described in the succeeding chapter.

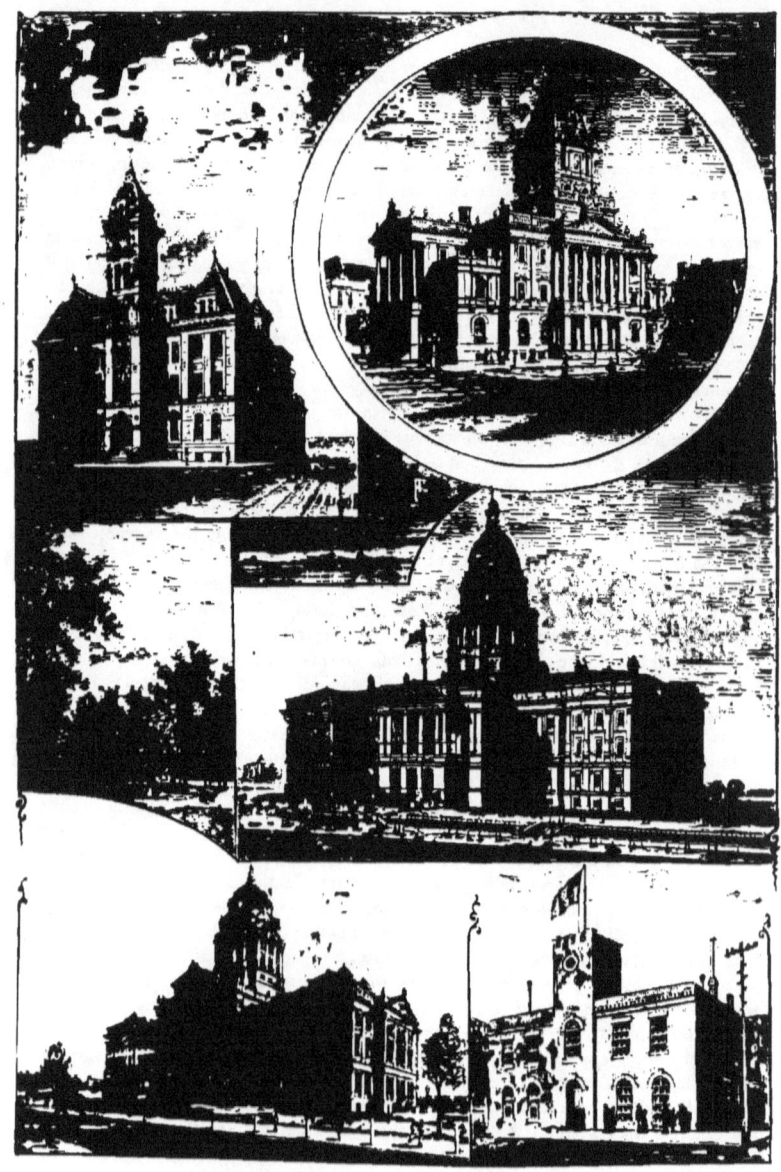

VIEW OF PUBLIC BUILDINGS, DENVER.

CITY HALL.
ARAPAHO COUNTY COURT-HOUSE.
NEW POST-OFFICE.
NEW CAPITOL BUILDING.
U. S. BRANCH MINT.

FROM DENVER TO PUEBLO.

> **DENVER,**
> Capital of Colorado.
>
> Population, 110,000.
> Elevation, 5,195 feet.

There are only a few cities in the world that please at first sight. Denver is one of this favored few. The liking one gets for Boston, Philadelphia or London is an acquired taste, but one falls in love at once with Paris, Denver or San Francisco. It does not follow that because the cities mentioned are immediately pleasing, they must of necessity resemble each other, any more than that a peach, an apple or an orange should have a similar flavor. We like the fruit and we like the cities without having to learn to like them, but not for the same reasons. One feels a sense of exhilaration in the atmosphere of Denver. The grand view of the Snowy Range of mountains to the north and west and the broad expanse of horizon-bounded plains to the east and south exalt the spirits, the bland but bracing breezes cool the fevered pulse and the abundant oxygen of the air thrills one like a draught of effervescing champagne. A beautiful city, beautifully situated is Denver, with broad tree-shaded streets, along each side of which flow streams of sparkling water, necessary to the growth of vegetation in a country where the annual rainfall is less than fifteen inches, with public buildings of massive proportions and attractive architecture, with residences erected in accordance with the canons of good taste, with innumerable lawns of shaven grass, ornamented with shrubs and flowers, with charming suburbs and an outlying country, studded with fertile farms and flowering or fruiting orchards, peace is within her dwellings and plenty within her palaces. Denver has now seventeen railroads, two cable roads constructed, three motor lines, one circle or belt railroad, sweeping around the southwestern segment of the city's circumference, and thirty-two miles of first-class street railway, which is being rapidly replaced by the extension of the cable road. The town is lighted by gas and electricity, has paid fire and police departments, and obtains its water from mountain sources by means of Holly works, and from over 300 artesian wells varying in depth from 350 to 1,600 feet. The public buildings, exclusive of churches and schools, cost $4,000,000. The real estate belonging to the city is worth $2,000,000, the bonded debt is only $400,000 and the assessed valuation of Denver is $37,500,000. The commerce of Denver is now annually not less than one hundred millions of dollars. Denver is situated at the junction of Cherry Creek and the Platte River, and in addition to being the Capital of the State, is the county seat of Arapahoe County. All the railroads which enter Denver land their passengers at the Union Depot, a massive and handsome edifice of native stone. The street leading from the main entrance of the station up town is Seventeenth Street, and on this just outside of the depot park is situated the central station of the City Street Railway Co. All the main lines of cars centre here, and should the traveler wish to reach some place in town by this means, a word of inquiry of the agent in the waiting room will elicit full information as to which line to take. The street and

FIRST CONGREGATIONAL CHURCH, DENVER.

cable cars pass directly by the leading hotels and radiate to all points of the city. On the town-ward side of the Union Depot are the carriage stands, and if arrangements for transportation have not already been made on the train, with the carriage company's agent, before reaching the city, a carriage can be engaged here. Prices are regulated by ordinance and extortion prohibited by law. There are many objects of interest to see in Denver : The smelters, the public buildings, the Grand Opera House—which is the handsomest in the world with the sole exception of the Grand Opera House in Paris,—the system of irrigation, the magnificent private

ARAPAHOE COUNTY COURT HOUSE, DENVER.

residences, the homes of mining princes and cattle barons, the lovely suburbs and the United States Military Post. The hotel accommodations of Denver are probably the most complete of any city of its population in the country. There are six first-class hotels provided with all modern improvements, to say nothing of some forty odd less pretentious ones. A day, or better two days, can be profitably spent in Denver, and then refreshed and rested from the long ride across the plains from the Missouri River or beyond, the tourist is ready to resume his trans-continental journey. If he wishes to behold the wonders of nature and to get a familiar acquaintance with the grandeur of the mountains, he will take the Denver & Rio Grande Railroad, which by universal acclaim has been designated "The Scenic Line of the World."

Seated in a comfortable car, whose large windows give an excellent outlook on the scenery, the traveler is ready and anxious to be off. The busy Union Depot may amuse him for a moment, but anticipation of the wonders in store makes him impatient of delay. Soon the conductor gives the signal to the engineer, the inevitable late passenger is seen chasing the rear end of the Pullman out of the depot, and whether he catches it or not, one thing is assured, the journey to the Pacific Coast has begun, and from this time on the eye and mind will both find plenty to do in noting and recording· Nature's most marvelous works. The first stop is made at

Burnham. The station for the suburb of West Denver and the site of the great shops of the Denver & Rio Grande Railroad. The buildings of the machine shops cover an area of five acres and were erected at a cost of $300,000. (Distance from Denver, 2 miles.)

Overland Park is a pleasant suburb to the southwest of Denver, and is supplied with one of the best race courses in the west. It is a fashionable resort

COLORADO'S STATE CAPITOL BUILDING, DENVER.

and connected with Denver by the suburban train service of the Denver & Rio Grande Railroad.

Petersburg is a small town surrounded by farms, market gardens and plats laid out as additions to Denver. (Distance from Denver, 8 miles.) To the west, 2½ miles distant, lies the United States Military Post.

Military Post. A ten company post of United States troops has been here established, and has become the centre of great interest. The quarters are elegant and substantial, consisting of handsome brick edifices. The parade ground is ample in proportions, and no expense has been spared to make this Post a model of its kind. The military band gives frequent concerts, and the citizens of Denver take great interest in and make frequent excursions to the Post. The Denver & Rio Grande Railroad has established a very complete suburban train service for the accommodation of the Post, and the public, which is largely patronized.

Littleton is prettily situated on the east bank of the Platte River, is the centre of a good agricultural country, and is destined to be the location of the suburban residences of many of Denver's best citizens. Already an adequate suburban train service has been inaugurated for the convenience of persons having country homes at this delightful spot. (Population, 300. Distance from Denver, 10 miles. Elevation, 5,372 feet.)

Acequia. A small station for the accommodation of ranchmen. Here the High Line Canal, one of those great irrigating ditches characteristic of Colorado, crosses the track and takes its winding way to the northeast over the rolling plains, having under its fertilizing power at least a hundred thousand acres of otherwise arid land. (Population, nominal. Distance from Denver, 17 miles. Elevation, 5,530 feet.)

Sedalia. A little village. Home market and post office for cattle growers and ranchmen. (Population, 100. Distance from Denver, 25 miles. Elevation, 5,835 feet.)

Castle Rock. The town takes its name from a peculiar upthrust of rock on the summit of a conical hill, resembling, in the distance, an old martelle tower, and nearer by an irregular pentagonal structure. Under the shadow of this hill and surmounting tower lies the town, which is a pretty village and the county seat of Douglas County. Fine quarries of red sandstone are worked here, and pastoral industries contribute to the prosperity of the town. (Population, 300. Distance from Denver, 33 miles. Elevation, 6,219 feet.)

Douglas. A station near which are stone quarries and grazing lands. (Population, nominal. Distance from Denver, 35 miles. Elevation, 6,323 feet.)

Between Douglas and Palmer Lake are the small stations of Glade, Larkspur and Greeland.

Perry Park is reached by stage from Larkspur station. This park abounds in curious formations of red sandstone; is watered by sparkling brooks and is destined to become one of the most popular resorts near Denver.

PALMER LAKE.
Health and Pleasure Resort.
Population, 150.
Distance from Denver, 52 miles.
Elevation, 7,237 feet.
Eating Station.

As the train rolls into the station the traveler sees to his left a beautiful little lake cradled in the hills. Along the shore has been placed a handsome cut stone embankment, and a neat and tasteful boat house has been erected and well stocked with boats. The lake is a natural body of water, though the fact that a fountain plays in its centre, casting a jet of water to the height of 80 feet, leads many to suppose that it is entirely artificial. Palmer Lake in addition to being a place of great beauty, is a natural curiosity, poised as it is, exactly on the summit of the "divide," a spur of the outlying range of the Rockies extending eastward into the great plains and from the crest of this summit the waters divide flowing northward into the Platte, which empties into the Missouri, and southward into the Arkansas as it wends its way to the Mississippi. Red roofed picturesque cottages nestle here and there among the hills, gayly painted boats float gracefully upon the bright blue waters, and on either hand rugged peaks, pine clad and broken by castellated rocks, rise into a sky whose cerulean hue is reflected in the placid waters of the lake. Excellent hotel and livery establishments furnish good accommodations for sojourners.

Glen Park, an assembly ground modeled after the famous Chautauqua, and destined to become equally as popular in the West as its prototype in the

PALMER LAKE.

East, is only half a mile beyond Palmer Lake. Objects of natural interest are abundant and the walks and drives to Glen D'Eau, Bellview Point, Ben Lomond, the Arched Rocks and the cañons and glens adjacent afford material for enjoyment in the seeing and for many pleasant memories. One hundred and fifty acres are comprised in the town site. The Park is at the foot of the Rocky Mountain Range, and is sheltered at the rear by a towering cliff 2,000 feet high, and on the two sides by small spurs of the range. A noble growth of large pines is scattered over the Park. A skillful landscape engineer has taken advantage of every natural beauty and studied the best topographical effect, in laying out the streets, parks, reservoirs, drives, walks, trails and lookout points. It is a spot that must be seen to be appreciated, and every visitor, whose opinion has been learned, has come away captivated. There are building sites for all tastes. Some have a grand lookout, taking in a sweep of the valley for a distance of 50 miles, with the fountain in Palmer Lake and the beautiful lake itself in full view. Elephant Rock, Table Mountain, the town of Monument, the railroad trains from both ways for over half an hour before reaching the station can be seen. Others have pretty vistas, partly hidden by the pine branches, promises, so to speak of grand views, but not so ambitious as the first. Still others are sylvan nooks where the shades are deepest and the murmur of the cool waters of the babbling brooks makes music forever.

Monument. The five miles ride from Palmer Lake to Monument is interesting. On the left are giant upthrusts of brilliant red rocks castelated in shape and reaching an altitude of two and three hundred feet. The town takes its name from the creek which flows near, and the creek is so designated from the curious monumental forms of rock along its course. To the right is the Front Range of the Rockies, which the road parallels from Denver to Pueblo, and near the centre of this stretch of one hundred and twenty miles, stands Pike's Peak. Agriculture and pastoral industries are tributary to Monument. (Population, 200. Distance from Denver, 56 miles. Elevation, 6,974 feet.)

Two miles beyond is Borst, and four miles further Husted, both mere side tracks for convenient shipping of cattle and produce.

Monument Park is reached by private conveyance from Edgerton Station—distance from Denver, 67 miles. This valley is quite remarkable for the very fantastic forms into which the action of air and water through long reaches of time, have worn the sandstone rocks, forming grotesque groups of figures that very generally keep their broad brimmed sombreros, formed of iron stained cap-rock. Visitors to Monument Park obtain a fine view of Pike's Peak and Cheyenne Mountain Range. A hotel in the Park is open at all times for the accommodation of guests, and can furnish saddle-horses and carriages on premises. The grotesque groups of figures into which the cream-colored sandstone rocks have been worn, some of them resembling human forms and have been given quaint, descriptive titles, viz.: Dutch Wedding, Quaker Meeting, Lone Sentinel, Dutch Parliament, Vulcan's Anvil and Workshop, Romeo and Juliet, Necropolis or Silent City, The Duchess, Mother Judy and Colonnade; all of these and many others too numerous to mention are within easy walking distance to "The Pines." The Park is a favorite resort and has comfortable accommodations for guests. (Population nominal. Distance from Denver, 67 miles. Elevation, 6,354 feet.)

DENVER & RIO GRANDE RAILROAD DEPOT AND PIKE'S PEAK.

COLORADO SPRINGS,

Residence City and Health Resort.

Population, 10,000.

Distance from Denver, 75 miles.

Elevation, 5,982 feet.

Many of the most influential business men of Colorado have their residence in Colorado Springs. No more delightful home city can be found than this. Mansions and cottages of the highest architectural beauty abound, and the society is composed of cultivated and wealthy people.

The town was originally laid out as a health resort, and while it still maintains its superiority in this respect, has grown beyond that single characteristic, and is now a thriving commercial place, in addition to being a favorite residence city. The town is sheltered on the west by the range of mountains with Pike's Peak in the centre, on the east by bluffs, on the north by the spur of the mountains called the "Divide," and on the southwest by Cheyenne Mountain. The streets are unusually wide, one hundred feet, and the avenues are 160 feet broad. Trees line both sides of the streets, and on Nevada avenue, the central street of the city, there are six rows of trees, two on each side and two down the centre. Water for irrigation is brought into the town by means of a winding canal, and cold, clear water, for domestic uses, is conducted from mountain sources in iron pipes. The pressure is such that no fire engines are necessary, the water being forced from hydrants to the tops of the tallest buildings. Monument Creek flows west of the town, and the Fontaine qui Bouille to the south, where the two streams form a junction. The scenery around Colorado Springs is of a very interesting and attractive character. The hotels of Colorado Springs are noted for their excellence; special attention being paid to the entertainment of tourists. There are ample accommodations and of different grades to suit all tastes and pockets. The Denver & Rio Grande Railroad has a very handsome stone depot, erected in accordance with good taste and correct architecture. The plains to the east and the mountains to the west give unlimited variety. Cheyenne Cañon, Austin's Bluffs, Crystal Park, Cameron's Cone, Monument Park and Manitou, with its environs, are all within the radius of nine miles.

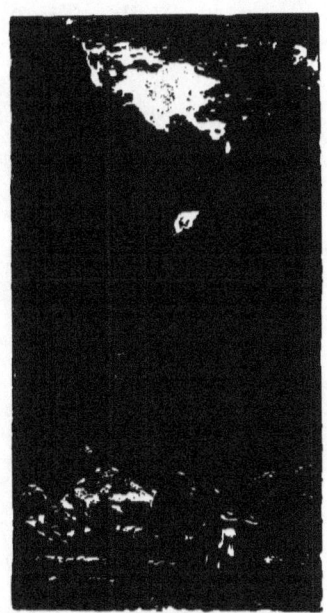

RAINBOW FALLS.

Cheyenne Mountain. It is impossible to contemplate the grandeur of Cheyennne's bold outline and great massiveness, and to become in the least familiar with its ever-varying play of light and shadow, without acknowledging the striking beauty of this noble mountain. From Colorado Springs a superb view of its front is seen. Looking at the mountain it will be observed that at almost the nearest point, in reality four miles distant, the base of the mountain is deeply cleft by two yawning chasms, the outer rocks of which present sharp, jagged points. These clefts are, respectively, the North and South Cheyenne Cañons. They certainly should be visited by every traveler who

has an eye for the beautiful. On the eastern side of Cheyenne Mountain, and accessible from South Cheyenne Cañon, is the grave of the well-known author and poet, "H.H." The direct road from Manitou takes the tourist a distance of eight miles, turns off to the southward from the road to Colorado Springs, on the top of the hill half a mile from the town; they can also be reached by making a detour of one and a half miles through Colorado Springs, and following the continuation of Nevada avenue to the southward. Either road is pleasant, and the drive or ride is one replete with interest, and abounding in attractive scenery.

Colorado City. This town, once the seat of the state capital, is two miles west of Colorado Springs, on the Manitou branch of the Denver & Rio Grande Railroad. The town gave promise of becoming an important city during the early days of its capitalship, but when the state government was removed to Denver, Colorado City languished, and soon sunk to the condition of a mere rural hamlet. This state of affairs lasted for a long series of years until property held only a nominal value. Recently, however, the advent of another railroad, the erection of shops, and the introduction of new industries awoke the town, and an era of great growth and improvement has set in. Holders of property have become rich, and the "old town," as it was called, is one of the most thriving in the state. (Population, 1,800. Distance from Denver, 78 miles. Elevation, 6,110 feet.)

> **MANITOU,**
> Watering Place,
> Mineral Springs
> and Health Resort.
>
> Population, 1,000.
>
> Distance from Denver, 81 miles.
>
> Elevation, 6,324 feet.

The one resort of all the West is certainly Manitou. The attractions of this watering place have secured for it fame, and fame secures for it largely increasing patronage each year. No resort has had a more rapid growth than this, and none has more truly deserved its prosperity. There are more places of extraordinary interest to visit in the vicinity of Manitou than can be found contiguous to any other resort in the world. It is situated six miles from Colorado Springs, immediately at the foot of Pike's Peak. Here are the famous effervescent soda and iron springs which in an early day gave the name of "Springs" to the town of Colorado Springs. A branch of the Denver & Rio Grande Railroad unites the two places, over which trains run daily with sufficient frequency to accommodate the most exacting. There are a thousand ways in which to enjoy oneself in Manitou. A favorite pleasure is that of riding. The saddle horses are excellent. Comfortable saddles for ladies and well trained horses are furnished by all the livery stables at reasonable prices. A burro (donkey) brigade is a feature for the special benefit of the children, a careful guide taking the little ones for a ride every morning. Carriage riding and excursions on foot are excellent means of diversion. Following is a partial list of places of interest near Manitou, with the distance in miles from town attached:

Manitou Grand Caverns	1
Cave of the Winds	1
Ute Pass and Rainbow Falls	1½
Red Cañon	3
Crystal Park	3
Garden of the Gods	3
Glen Eyrie	5
Monument Park, by trail	7½
" " by carriage	9

MANITOU SPRINGS AND PIKE'S PEAK.

Seven Lakes, by horse trail............................. 9
      "   "   by carriage road......................... 25
North Cheyenne Cañon.......... .... 8½
South Cheyenne Cañon 9
Summit of Pike's Peak................................... 12

In addition to these well-known localities there are scores of cañons, caves, water-falls and charming nooks which the sojourner for health or pleasure can seek out for himself. The village is thronged with visitors throughout the summer months; it is somewhat cooler and less dry than Colorado Springs in the summer, and warmer in winter. The springs all contain more or less soda and some iron. They are peculiarly adapted for the dyspepsia of the consumptive, and the Ute Iron Spring is especially remarkable for its blood-making qualities. For the pleasure-seeker and the invalid, Manitou is one of the most satisfactory resorts in the State. During the season the hotels are filled with guests from all parts of the Union. Society is represented by many of its best people, the evenings are made merry with hops and social gatherings, and the days delightful with drives and rides and walks among the myriad of attractions this place affords.

PIKE'S PEAK.

Colorado's Landmark.

Elevation,

14,147 feet.

Before Colorado had acquired a name, Pike's Peak was the landmark of the Indian, the trapper and the explorer. In later times it was the beacon by which the adventurous gold hunters steered their prairie schooners into the wonderful and mysterious west; now it has become the goal of those

THE SEVEN FALLS, CHEYENNE CAÑON.

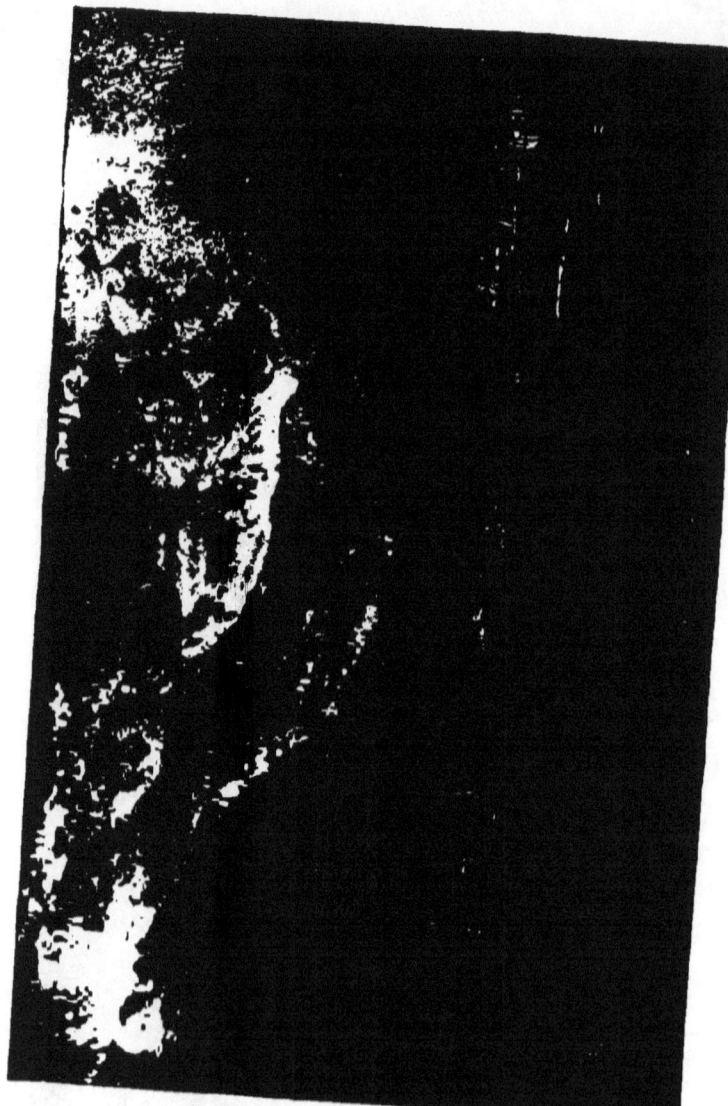

GLIMPSE OF PIKE'S PEAK.

in search of the grand and beautiful in Nature, the enjoyments of an attractive summer resort, or the restoration of impaired health. The mountain is one of great beauty, easy of ascent and never entirely discrowned of snow. To any one accustomed to mountain climbing no guide is required in making the ascent of Pike's Peak, as the trail is good and well-defined, and there is a government station on the summit, where visitors can obtain food and shelter. Three members of the United States Signal Service Corps live on the Peak all the year round, and are in constant telegraphic communication with Colorado Springs and Washington. The telegraph poles for the most part closely follow the trail. At the end of the Ruxton Glen road is a toll gate, and here the ascent of the Peak begins. For three miles the trail closely follows Ruxton Creek, generally at an elevation of two or three hundred feet above it; the sides of the Glen are clothed with beautiful pines and spruces. Close to the creek the familiar bunches of red berries characteristic of the mountain ash may here and there be greeted, as old friends in a strange land. Some very pretty falls are passed on the way, two of which are named respectively, the Shelter and the Minnehaha. Stupendous granite boulders are in places piled up in chaotic confusion over the stream, frequently hiding it from view. Two prominent ones are plainly visible from Manitou, and are appropriately named Gog and Magog. One of the most charming features during the first three miles of the ascent is the opportunity afforded for exquisite views of the world below, on looking back through the pine trees with the far-stretching plains glowing in the sun and forming a golden horizon. It goes without saying that the view from the summit is grand beyond description. Within the current year the visitor to Manitou will in all probability be able to continue his journey by rail from the latter point straight up Pike's Peak to its summit, a height of 14,147 feet above the level of the sea. The Manitou & Pike's Peak Railway Company have under construction a line to the summit similar to the cog rail line on Mount Washington. This will be the most direct route, estimated length, 8 miles.

Fountain. A pretty little town on the Fontaine qui Bouille Creek, fourteen miles south of Colorado Springs. The town has taken a new growth within the past year, and being surrounded by a good grazing and agricultural country, has a fair prospect of permanent improvement. (Population, 200. Distance from Denver, 88 miles. Elevation, 5,568 feet.)

There are between Fountain and Pueblo, side-track stations as follows: Buttes, Wigwam, Piñon, Cactus and Nada. These places are useful to the railroad and convenient for the residents of the surrounding country, but they possess little or no interest for the traveler. All the way from Denver to Pueblo the traveler has the Front Range of mountains on his right, to the west, while on his left are the great plains. Below Colorado Springs the country is very fertile, and good crops are grown wherever water for irrigation can be procured. Water is scarce, however, and only a small part of the land is under cultivation.

"The Pittsburgh of the West" is a title often conferred on Pueblo, and it is the name which pleases its citizens best, and which comes the nearest to expressing the salient characteristics of the town. It is a live city, full of enterprise and push, and it has been favored by Nature, both in the matter of its immediate situation and of its surroundings. Plenty of coal is found not fifty miles away, iron ore is not more distant, and on the mesa, just south of the town, is Bessamer, the site of

---

PUEBLO,
Commercial and Manufacturing City.
Population, 25,000.
Distance from Denver, 120 miles.
Elevation, 4,067 feet.
Dining Station.

IN THE GARDEN OF THE GODS.

the Colorado Coal and Iron Works, one of the largest plants of this character in the world. There are also many great smelters for the reduction of gold and silver ores together with a large number of manufactories, planing mills, flouring mills, machine shops, etc. The city of Pueblo, is surrounded by great stretches of rich agricultural land, which in places here and there is under a high state of cultivation. But it is only here and there that cultivation shows its elevating work. Tourists wonder at this, and cannot divine why, if the land is rich, it should lie fallow and uncultivated. The answer is easy to find. All this land is arid. Crops will not grow without water, and the rains of heaven are not half copious enough to promote the growth of vegetation. Where the land is watered by irrigation it is as fertile as the valley of the Nile, where it is not irrigated it is nearly as sterile as the desert of Sahara. This condition of affairs will not long remain. Storage reservoirs to conserve the winter and spring rainfall and snow deposits are in contemplation, also a series of great canals to be taken from the Arkansas river to carry the water on to the waiting land. In the mean time this uncultivated country, which appears so barren, supports tens of thousands of sheep and cattle. The short, dry crisp, curled buffalo grass, which looks about as succulent as shavings, actually contains great nutritive qualities, and if cattle or sheep can get enough of it they grow fat and command the highest price in the markets. Pastoral and agricultural interests contribute to Pueblo's prosperity, five trunk lines of railroad centre here, and manufactories increase the business of the town. Many people of great wealth make Pueblo their home and do business here. Handsome mansions, pretty cottages, large business blocks, and fine stocks of all kinds of merchandise testify to the good taste and enterprise of Pueblo's citizens. It is admitted on all sides that this must of necessity become the leading manufacturing town between the Missouri river and the Pacific coast, and the manufacturers in the East who contemplate extending or removing their works, are now carefully studying the resources of Pueblo. Pueblo is well provided with hotels, one of them representing an expense of $250,000 in its erection. All grades of excellence can be found among the hostelries, and the traveler will find no difficulty in securing accommodations suited to his tastes. Through Pueblo, the traveler passes to reach Santa Fe, Española, Durango and Silverton on the south, Leadville, Glenwood Springs and Aspen on the northwest, or Cañon City, Salida, Gunnison, Montrose, Ouray, Grand Junction, Salt Lake City, and Ogden on the west, *en route* to San Francisco.

Parnassus Springs. A pleasant drive of twelve miles, southwest of Pueblo, takes us to Parnassus Springs, among the foot hills of the Greenhorn Mountains. These waters—muriated alkaline—have been tested with marked benefit, especially in cases characterized as gastric complaints.

Carlile Springs are situated twenty miles above Pueblo, on the Arkansas river. These purgative alkaline waters are as yet unimproved, but give good promise of becoming popular on account of their medicinal qualities.

Clark's Magnetic Mineral Spring. This celebrated spring in the suburbs of Pueblo, has recently been improved by the erection of a large bath house, fitted up with all the latest improvements and conveniences for bathing.

A DONKEY BRIGADE.

PUEBLO TO OGDEN.

ROM Denver to Pueblo, a distance of one hundred and twenty miles, the traveler has followed the Front Range of the Rocky Mountains and kept his course mainly to the south At Pueblo, however, he turns his face westward, and this will be his outlook, in the main, until he finds himself standing on the shore of the Pacific Ocean, watching the descent of the sun into the wilderness of waters. The country between Pueblo and Florence is fine agricultural land, being the bottoms of the Arkansas River, up whose course the railroad follows until Salida is reached, ninety-seven miles from Pueblo. Back from the river rise high buttes of sandstone worn into fantastic shapes by the action of the elements. Banded with a great variety of colors and dotted here and there by groups of pines, the scene is one of much interest and adds an element of variety to the journey, which is exceedingly grateful to the traveler. The river bottoms are irrigated by means of ditches taken from the river, and the result is crops of marvelous growth and yield. One interesting and peculiar feature is the frequent occurrence of the ancient Egyptian water wheels suspended in the current of the Arkansas. This method of securing water for irrigation is rarely observed in Colorado. This valley of the Arkansas is also a good fruit country, and grapes and apples grow in abundance and of fine quality.

Florence. This town is in the centre of the coal oil fields of Colorado. Glancing from the car window the traveler will here see the tall derricks of the well machinery and the tanks for storing, together with the tank cars for transporting the oil. There are between thirty and forty wells already in operation and more are being sunk. The oil is used for lubrication and fuel, and gives the best of satisfaction. Florence is a growing town and a pretty one, surrounded by an attractive country. (Population, 1,000. Distance from Denver, 152 miles. Elevation, 5,199 feet.)

Coal Creek Branch. A branch line of the Denver & Rio Grande Railroad runs from Florence to Coal Creek, a distance of six miles, where excellent and extensive coal mines are in operation. This line is one of great commercial importance, opening one of the most extensive coal fields in the state.

Coal Creek is at the terminus of this branch of the line. It is well supplied with stores and shops of all kinds and does a thriving business. (Population, 1,500. Distance from Denver, 155 miles. Elevation, 5,360 feet.)

CANON CITY,

Health and Pleasure Resort.

Business Centre.

This city is rightly named, for it stands at the entrance to the greatest cañon penetrated by any railroad. The Grand Cañon of the Arkansas is acknowledged by a universal consensus of opinion to be one of the great wonders of the world. The Arkansas River, which rises in Fremont Park, one hundred and seventy-five miles to the northwest of Cañon City, here breaks its way through the Front Range of mountains and enters upon its uneventful course to the Mississippi. The town is one of the oldest in Colorado, and is essentially a place of pleasant homes. It is the county seat of Fremont County, and the seat of the State Penitentiary. Its warm and equable climate makes it a favorite resort for invalids. In addition to its pleasant climate it possesses valuable mineral springs, both hot and cold. The water of the cold springs is almost icy in temperature, and strongly impregnated with soda. The cold springs are situated just above the Penitentiary. The scenery round about Cañon City is exceedingly attractive. The drive of about twelve miles to the brink of the Royal Gorge and the view of that wonderful chasm from the top, which can there be obtained, are experiences never to be forgotten. The town and its contiguous country possess the finest orchards in the state, and the cultivation of fruit has become a leading industry. The city is well built, has handsome business blocks and comfortable and elegant residences. (Population, 2,500. Distance from Denver, 161 miles. Elevation, 5,243 feet.)

The Hot Springs. Having left Cañon City and traversed a mile to the westward the traveler will observe to his left a picturesque, many gabled building, across the river, a rustic foot bridge leading thereto.

GRAPE CREEK CAÑON.

This is the Royal Gorge hotel situated at the Hot Springs. The hotel has excellent accommodations for guests and is a favorite resort for health and pleasure seekers. The springs are recommended by physicians as excellent in cases of cutaneous and blood diseases. Prof. Loew's analysis of the waters is as follows:

	Grains in a Gallon of Water Temperature 104 deg. Fah.
Chloride of Sodium	18.2
Sulphate of Soda	79.3
Carbonate of Soda	73.2
Carbonate of Lime	33.5
Carbonate of Magnesia	12.8
Lithia	Trace.
	217.0

THE ROYAL GORGE.

Baths have been provided at the hotel and are supplied with all of the modern conveniences.

Silver Cliff Branch. This branch, 33 miles in length, which turns to the left just as the train enters the Grand Cañon, two miles above Cañon City, has its terminus at West Cliff. It passes through most charming scenery and enters an exceedingly fertile country, the Wet Mountain Valley surrounding the terminal station. Its greatest claim to scenic attraction is the fact that it passes through a cañon only less grand than that of the Arkansas.

Grape Creek Cañon. Among the many remarkable cañons for which the State of Colorado is famous, there is probably none which presents more attractions to the lover of nature, or which combines the sublime with the beautiful more perfectly, than that of Grape Creek. This beautiful stream takes its rise among the lofty and almost inaccessible peaks of the Sangre de Cristo Range, and flowing nearly northward, waters in its course the beautiful and fertile Wet Mountain Valley; then passing near the famous Silver Cliff mining camp it continues its tortuous course in an easterly direction until it enters the Arkansas River about a mile above Cañon City, just where the river leaves the Grand Cañon, after its terrific conflict with the granite cliffs, and tossing its foam crests high in the air, makes its last triumphant exit from the mountains. The walls of this cañon present a splendid study for the geologist, as piled up in many places over a thousand feet in nearly vertical height, they exhibit the various formations of primary rock in a striking and peculiar manner. The entrance to the cañon for over a mile follows the windings of the clear flowing creek, with gently sloping hills on either side covered with low spruce and piñon, and with grass plats and brilliant flowers, in season, far up their slopes, and the Spanish lance and bush cactus present their bristling points wherever a little soil affords them sustenance. To examine this cañon thoroughly a carriage or saddle-horses should be taken from Cañon City, but as the train ascent of the grades must be made slowly, a very satisfactory view can be gained from the cars in passing.

West Cliff. This town is beautifully situated in the Wet Mountain Valley, surrounded by a fine grazing and agricultural country. The view is a grand one, lofty mountains bounding the entire circle of the horizon. A mile from the station is Silver Cliff, which after the discovery of the Racine Boy mine, was the centre of a tremendous rush of miners, resulting in several other great discoveries, but the large mines were few in number and the prospectors left for other fields. The good mines are still productive and add their quota to the prosperity of the valley. West Cliff is the shipping point for Silver Cliff and Rosita, being the railroad station. (Population, 800. Distance from Denver, 194 miles. Elevation, 7,864 feet.)

ROYAL GORGE.

Distance from Denver, 163 miles.

Greatest Height of Walls, 2,627 feet.

Length, 7 miles.

Just beyond Cañon City the railway enters the Grand Cañon of the Arkansas, the narrowest portion of which is known as the Royal Gorge. When first examined it seemed impossible that a railway could ever be constructed through this stupendous cañon to Leadville and the west. There was scarcely room for the river alone, and granite ledges blocked the path with their mighty bulk. In time, however, these obsructions were blasted away, a road-bed closely following the contour of the cliffs was made, and to-day the cañon is a well-used thoroughfare. But its grandeur still remains. After entering its

MARSHALL PASS—EASTERN SLOPE.

THE ROYAL GORGE

depths, the train moves slowly along the side of the Arkansas, and around projecting shoulders of dark-hued granite, deeper and deeper into the heart of the range. The crested crags grow higher, the river madly foams along its rocky bed, and anon the way becomes a mere fissure through the heights. Far above the road the sky forms a deep blue arch of light; but in the Gorge hang dark and sombre shades which the sun's rays have never penetrated. The place is a measureless gulf of air with solid walls on either side. Here the granite cliffs are a thousand feet high, smooth and unbroken by tree or shrub; and there a pinnacle soars skyward for thrice that distance. No flowers grow, and the birds care not to penetrate the solitudes. The river, sombre and swift, breaks the awful stillness with its roar. Soon the cleft becomes still more narrow, the treeless cliffs higher, the river closer confined, and where a long iron bridge hangs suspended from the smooth walls, the grandest portion of the cañon is reached. Man becomes dwarfed and dumb in the sublime scene, and Nature exhibits the power she possesses. The crags menacingly rear their heads above the daring intruders, and the place is like the entrance to some infernal region. Escaping from the Gorge, the narrow valley of the upper Arkansas is traversed, with the striking serrated peaks of the Sangre de Cristo close at hand on the west, until Salida is reached.

During the summer season an open observation car is attached to each through train while traversing the Grand Cañon and the Black Cañon, thus affording the traveler the best opportunity of seeing these wonders of nature. There are a number of stations between Cañon City and Salida, but none of them are of special interest to the tourist, except that fishing and hunting can be found in the immediate vicinity of any of them.

Parkdale. At this little station the observation car is detached from the west bound and attached to the east bound train. From this open car the tourist can obtain an unobstructed view of the grandeurs of the Royal Gorge, and is in service during most of the year; being discontinued only during the most inclement months of winter.

Beautiful Mountain View. Emerging from the cañon, a most beautiful mountain view is obtained; to the left stretch the serrated summits of the Sangre de Cristo Range, while to the front and right are the towering peaks of the Collegiate Mountains.

Wellsville Hot Springs are on our left across the Arkansas River, six miles before Salida is reached. Here is a natural warm plunge bath, the waters of which are strongly impregnated with medicinal qualities. The Wellsville Springs are a favorite resort, and are made the objective point for many pleasant excursion parties.

SALIDA.
Health and Pleasure Resort and Business Centre.
Population, 3,000
Distance from Denver, 217 miles.
Elevation, 7,049 feet.
Eating Station.

This prosperous town is situated on the right bank of the Arkansas River, at the junction of the Leadville and Aspen branch of the Denver & Rio Grande Railroad, with the main line to Salt Lake and Ogden. The view of the mountains from Salida is especially grand. The Collegiate Range rises to the west with Yale, Harvard and Princeton Peaks in plain view crowned with perpetual snow, while to the south stands the Sangre de Cristo Range, and in the southwest tower Ouray and Shaveno. The beauty of its situation, the near proximity to hot medicinal springs, the wonderful salubrity of its climate, make Salida an extremely popular health and pleasure resort. Tributary to the town are mines of copper, silver, gold, iron and coal, great quantities of charcoal are burned near Salida, and the agricultural and pastoral interest are of great extent.

Poncha. This little town, five miles west of Salida, is the station for Poncha Hot Springs and the junction of the Monarch Branch with the main line It is really a suburb of Salida, and is connected with that town by a beautiful boulevard, which is one of the pleasantest of drives.

Monarch Branch. From Poncha this branch runs into a rich mining country, its terminus is Monarch, a prosperous mining town, 237 miles from Denver and 11 miles from Poncha. The intermediate stations on the line are Maysville and Garfield. Mining is the chief industry.

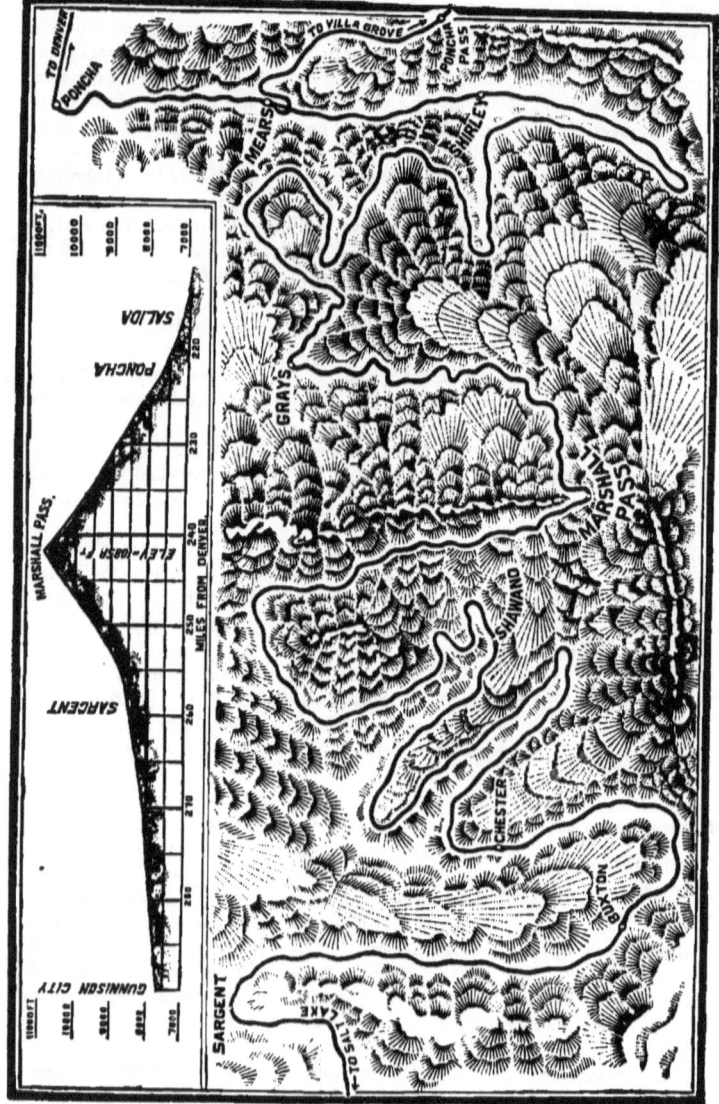

ALIGNMENT OF THE DENVER & RIO GRANDE RAILROAD OVER MARSHALL PASS.

PONCHA SPRINGS,

Hot Springs,
Watering Place,
and Health Resort.

Distance from Denver-
221 miles.

Elevation, 7,480 feet.

As a resort for invalids, Poncha Springs offers superior inducements, especially to those suffering from chronic troubles. The sick get well here in less time and with less medicine than in any other sanitarium outside of Colorado. The return to health here is made radically permanent. A great variety of diseases are cured by the peculiar earth-heated and earth-medicated waters and an intelligent system of baths. The effect on the sick is wonderfully beneficial, corollating a specific energy with the climate and pure atmosphere and the very feeble are enabled to tolerate much hotter baths than in damper or lower altitudes, and secure correspondingly greater results. The analysis of the Poncha Hot Springs corresponds almost exactly with the waters of the Hot Springs in Arkansas. The temperature of the various Arkansas Hot Springs varies from 90 to 175°, that of the Poncha Springs varies from 90 to 185° Fahrenheit. The water is as clear as crystal and perfectly odorless and tasteless. It quenches thirst whether cold or hot, and does not disturb the stomach in any manner. There are one hundred of these Hot Springs, all flowing from a great field of *tufa*, the natural precipitation of ages of loss of temperature from contact with the atmosphere and chemically the same as the *tufa* of the Arkansas Hot Springs. The springs have a capacity large enough to bathe 40,000 persons daily. Commodious bath-houses have been erected and competent physicians are in attendance. The following is an analysis of the Poncha Hot Springs:

Silicic Acid	32.73	Organic Matter	6.24
Sesqui-oxide of Iron	1.27	Water	1.72
Alumina	5.20	Sulphuric Acid	4.46
Lime	20.00	Potash	2.08
Magnesia	.74	Soda	1.00
Cholorine	.06	Iodine	1.50
Carbonic Acid Gas	22.50	Bromine	1.50

The waters are said to be a sure cure for rheumatism and all blood and skin diseases, and catarrhal affections.

Poncha Pass. After leaving Poncha Station the railroad begins to climb the mountains, and makes its entry into Marshall Pass by way of Poncha Pass. As the train makes a long curve around the side of a great hill, about two miles above the town of Poncha, the tourist can see the Hot Springs on the side of the opposite hill to the left, a deep gorge intervening, at the bottom of which flows a clear mountain stream. The scenery here is wild and beautiful, and the interest increases with each mile of the ascent.

Mears Junction. This little station, 227 miles from Denver, in the heart of the hills, is the junction of the San Luis branch with the main line, and from this point the real ascent of Marshall Pass begins.

San Luis Branch. This branch extends from Mears Junction to Villa Grove and Hot Springs, the latter point being the terminus of the line. The intervening stations are Round Hill and Davenport.

Villa Grove. This town is situated at the northern extremity of the great San Luis Valley, and is surrounded by a rich agricultural country. There are many good mines of gold, silver and coal, in the near vicinity. (Population, 200. Distance from Denver, 247 miles. Elevation, 7,971 feet.)

SANGRE DE CRISTO RANGE—FROM MARSHALL PASS.

MARSHALL PASS.

Railroading
Among the Clouds.
A Marvel
of Engineering Skill.
Elevation, 10,856 feet.

After leaving Mears Station the line advances by means of a series of curves absolutely bewildering, following the convolutions of the gulches. As the altitude grows greater, the view becomes less obstructed by mountain sides, and the eye roams over miles of cone-shaped summits. The timberless tops of towering ranges show him that he is among the heights and in a region familiar with the clouds. Then he beholds, stretching away to the left, the most perfect of all the Sierras. The sunlight falls with a white, transfiguring radiance upon the snow-crowned spires of the Sangre de Cristo Range. Their sharp and dazzling pyramids, which near at hand are

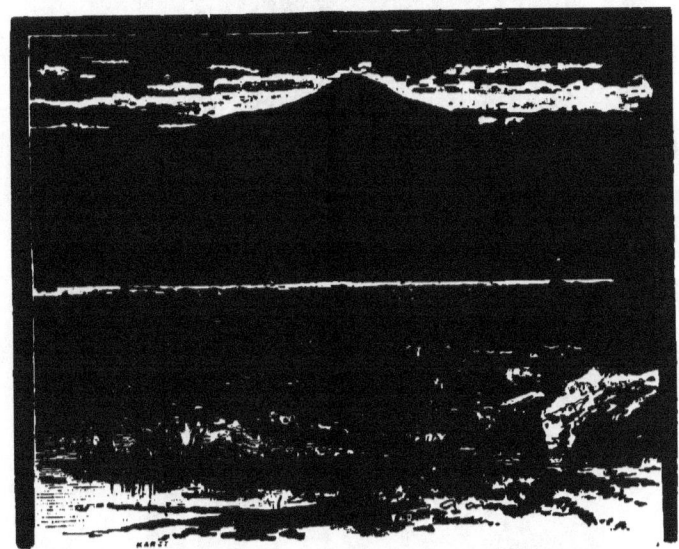

CRESTED BUTTE MOUNTAIN AND LAKE.

clearly defined, extend to the southward until cloud and sky and snowy peak commingle and form a vague and bewildering vision. To the right, towers the fire scarred front of old Ouray, gloomy and grand, solitary and forbidding. Ouray holds the pass, standing sentinel at the rocky gateway to the fertile Gunnison. Slowly the steeps are conquered until at last the train halts at the station, upon the summit of Marshall Pass. The awful silence of the storm-tossed granite ocean lies beneath. The traveler looks down upon four lines of road, terrace beyond terrace, the last so far below as to be quite indistinct to view. These are only loops of the almost spiral pathway of descent. Wonder at the triumphs of engineering skill is strangely mingled with the feelings of awe and admiration at the stupendous grandeur of the scene.

Marshall Pass Station. Is directly on the summit of the pass, and the track is enclosed by a large snow shed. Fine views can be obtained, however,

MARSHALL PASS—WESTERN SLOPE.

from the loop holes or from either end of the shed. The elevation is 10,852 feet above the sea. The descent begins, and the road winds around projecting headlands, on the verge of vast precipices, threads dark recesses where patches of light fall through leafy canopies upon the green slopes, follows the windings of the Tomichi, and later courses through cultivated meadows dotted with hay-stacks and small ranch houses. As the train rolls swiftly on, a backward glance gives the traveler a comprehensive idea of the vast heights overcome in the passage. The stations between Marshall Pass and Gunnison are as follows: Hillden, Shawano, Chester, Buxton, Sargent, Elko, Crookton, Doyle, Bonita, Parlin and Mounds. These stations are all small but situated in the midst of beautiful scenery.

The Waunita Hot Springs are situated eight miles from Parlin. The waters have long been famous for their great medicinal qualities, and they have been frequented by those suffering from ill health with the most surprising and gratifying results. Good accommodations have been provided for guests. The scenery surrounding the Springs is unsurpassed, and no pleasanter place can be found by the searcher after health or pleasure.

Tomichi Meadows. Beyond Parlin the line crosses a wide expanse of natural meadow land, through which meanders the beautiful Tomichi Creek.

GUNNISON.
Population, 2,500.
Distance from Denver, 290 miles.
Elevation, 7,683 feet.
Eating Station.

Gunnison is the county seat of Gunnison County, and is situated on the Gunnison River. From its central position in the great Gunnison Valley, it must of necessity always be the distributing point; and, therefore, its growth is assured as being coincident with that of the country in which it is situated. From Gunnison extends a branch of the Denver & Rio Grande Railroad up to Crested Butte, situated in the heart of a rich gold and silver mining country, and being the centre of the wonderful anthracite coal measures of the state. The town is beautifully situated and is in such close proximity to some of the most attractive scenery in the Rocky Mountains, that it has become a favorite objective point with tourists. The Gunnison River and its many confluent trout brooks offer fine sport for the fisherman, and the hills abound in game. The La Veta Hotel, the eating station for passengers, is one of the most magnificent in Colorado, having been erected at an expense of $225,000. It is elegantly furnished, and offers first class accommodations for the tourists who may wish to spend a few days or weeks here, hunting and fishing.

Crested Butte Branch. From Gunnison the Crested Butte branch of the road extends to the northwest to Crested Butte, a distance of twenty-eight miles. The line extends up the Gunnison River, which swarms with trout and is an extremely picturesque stream. The Elk Mountains are in plain view, and add grandeur to the scene. The intervening stations are Almont, Jack's Cabin and Glaciers.

Crested Butte. This pretty village is situated most delightfully among the mountains, one castellated peak directly opposite the town conferring the name it bears. This is the center of the most remarkable coal region yet discovered in Colorado, and abounding also in rich mines of gold and silver. At Crested Butte, just back of the village, is found abundant measures of exceedingly bituminous coal, which is mined largely and made into coke. Four miles north of the town anthracite coal, equal in every respect to the best found in Pennsylvania, is taken from the top of a mountain, and shipped all over Colorado and Utah. The fishing

GATE OF LADORE

and hunting in the mountain streams, and over the wooded hills, furnish abundant sport for the residents and tourists, and the rides and drives afford an almost infinite variety. (Population, 1,200. Distance from Denver, 318 miles. Elevation, 8,878 feet.)

Sapinero stands at the eastern entrance to the Black Cañon, and is beautifully situated on the banks of the Gunnison River. The town was named after a sub-chief among the Utes, who was regarded by the whites as a man of unusual intellectual and executive ability. In addition to commanding the entrance to the cañon, Sapinero is the junctional point for the Lake City extension of the line. (Population, 48. Distance from Denver, 316 miles. Elevation, 7,255 feet.)

Lake City Branch. This extension is thirty-six miles in length, and has its terminus at Lake City. The line turns to the left about a mile west of Sapinero, and passes through a remarkable cañon en route.

Lake Fork Cañon. This cañon is a most attractive bit of scenery. It is noted for its narrowness, and the height and grandeur of its walls. For thirteen miles the railroad winds through this tortuous chasm, the walls rising on each hand to a height varying from eight hundred to thirteen hundred feet. The river claims the right of way but the railroad also asserts its rights, and by the exercise of engineering skill has forced a passage. In many places the solid wall of granite has been blasted away, and from the fallen blocks a solid embankment constructed, upon which the rails have been laid. The Lake Fork is a rapid and tumultuous stream, abounding in rapids and presenting a most interesting, varied and exhilarating panorama to the eye. Emerging from the cañon and gaining a greater altitude, the view is one of magnificent extent and grandeur. Northward the peaks of the Elk Range form a long line of well-separated summits. Northeastward, the vista between nearer hills is filled with the clustered heights of the Continental Divide in the neighborhood of the Mount of the Holy Cross. Just below them confused elevations show where Marshall Pass carries its lofty avenue, and to the southward of that stretches the splendid, snow trimmed array of the Sangre de Cristo.

LAKE CITY.
Picturesque Mining Town.
Population, 1,500.
Distance from Denver, 352 miles.
Elevation, 8,604 feet.

This enterprising and thriving mining town stands in a little park at the junction of the Lake Fork of the Gunnison River with Hensen Creek, both typical mountain streams. A substantial and pretty town has been established. Mines of marvelous value surround the town, and the recent advent of the railroad has given new life and energy to all the commercial and speculative projects of the people. The development of her mining resources has been retarded during the past by the lack of transportation facilities, but this has only caused its stores of wealth to be held in abeyance for awhile before their coinage. Many another district, a few years ago thought equally profitless, has risen to become the scene of steady dividend making labor through the perfection of processes. It will not be long before, by like means, the reviving of Lake City's mines will occur, and enable her to catch up with her more fortunate sisters in the wide circle of the San Juan silver region. The romantic surroundings of this pretty town,—the lovely lakes from which it takes its characteristic name, the grand mountains and the grassy parks — have made it a favorite for the lovers of nature in the past, and will still attract them in the future. This is a paradise for a sportsman. Over these rolling uplands, among the aspen groves, upon the foot hills and along the

APPROACH TO THE BLACK CAÑON.

willow-bordered creek deer now throng, and even an occasional elk and antelope are to be seen. In the rocky fastnesses the bear and panther find refuge, and every little park is enlivened by the flitting forms of timid hares and the whirring escape of the grouse disturbed by our passing.

> **BLACK CANON**
> OF THE
> **GUNNISON.**
> Height of Walls,
> 2,500 feet.
> Length of Canon,
> 14 miles.

Beyond Gunnison, the railway traverses the valley of the same name, following the river closely, and encountering nothing but meadows and low, grayish cliffs. The Gunnison River abounds in fish, and is a great resort for the disciples of Isaac Walton. Soon, however, the channel, which the stream has worn, becomes narrower. The cliffs grow higher and steeper, the vegetation is less abundant, and suddenly the sunlight is cut off by broken summits, and directly after leaving Sapinero, where the observation car is attached, the Black Cañon holds us fast in its embrace. This gorge is grander, deeper, darker, and yet more beautiful than the one we have so lately penetrated. It is twice as long, has more verdure, and, although the walls are dark-hued enough to give the place its name, still they are of red sandstone in many places, and from their crevices and on their tops, shrubs, cedars and piñons grow in rich abundance. The river has a deep, sea-green color, and is followed to Cimarron Creek, up which the road continues, still through rocky depths, to open country beyond. The Black Cañon never tires, never becomes commonplace.

Chippeta Fall starts from a dizzy height, is dashed into fragments by lower terraces, and, tossed by the winds, reaches the river in fine white spray; there another cataract leaps clear of the walls, and thunders unbroken upon the ground beside us. In the cliffs are smaller streams, which trickle down and are lost in the river below. At times the cañon narrows, and is full of sharp curves, but again has long, wide stretches, which enable one to study the steep crags that tower heavenward two or three thousand feet.

Currecanti Needle, the most abrupt and isolated of these pinnacles, has all the grace and symmetry of a Cleopatra obelisk. It is red-hued from point to base, and stands like a grim sentinel, watchful of the cañon's solitudes. At the junction of the Gunnison and the Cimarron a bridge spans the gorge, from which the beauties of the cañon are seen at their best. Sombre shades prevail; the stream fills the space with its heavy roar, and the sunlight falls upon the topmost pines, but never reaches down the dark red walls. Huge boulders lie scattered about; fitful winds sweep down the deep clefts; Nature has created everything on a grand scale; detail is supplanted by magnificence, and the place is one appealing to our deepest feelings. It greets us as a thing of beauty, and will remain in our memory a joy forever. Long ago the Indians of this region built their council fires here. By secret paths, always guarded, they gained these fastnesses, and held their grave and somber meetings. The firelight danced across their swarthy faces to the cliffs encircling them. The red glow lit up with Rembrandt tints the massive walls, the surging streams and clinging vines. They may not have known the place had beauties, but they realized its isolation, and fearing nothing in their safe retreat, spoke boldly of their plans.

Cimarron. Is a most attractive little station, nestled among the gulches on the banks of sparkling Cimarron Creek. Here is a meal station, and here the observation car is detached. Sportsmen make headquarters at Cimarron, for the

CURRECANTI NEEDLE, BLACK CAÑON.

hills are full of game and the streams abound in trout. (Population, nominal. Distance from Denver, 331 miles. Elevation, 6,906 feet.)

Cimarron Cañon. Where Cimarron Creek empties into the Gunnison through a short cañon, the road leaves Black Cañon, which continues on with the larger stream, heightening in awfulness. Down there the fall of the river increases so rapidly that to follow it to the end, the railroad would emerge a thousand feet above the valley which it seeks, if a practicable grade should be kept, so the engineers have turned the road out to the valley through Cimarron Cañon, and in four or five miles a verdureless expanse is reached, and for hours the road traverses a region which is picturesque in its poverty and desolation; and in the summer the distant and sun-heated buttes, with the arid plains between, remind the traveler of the wastes of Arabia Petra

TROUT FISHING ON THE CIMARRON.

Cedar Divide is reached directly after emerging from Cimarron Cañon. From here the Uncompahgre Valley, its river, and the distant, picturesque peaks of the San Juan are within full sight of the traveler. Descending to the valley and following the river past Montrose, the Gunnison is again encountered at Delta.

MONTROSE.
Population, 1,500.
Distance from Denver, 353 miles.
Elevation, 5,811 feet.

This town can take just pride in the grandeur of its mountain view. Situated in the Uncompahgre Valley, Montrose is almost surrounded by mountains. The San Juan Mountains tower into the heavens to the south, captained by Mounts Sneffles and Uncompahgre, both over fourteen thousand feet high. Along the western horizon trend the Uncompahgre Peaks to where the Dolores joins the Grand River, a distance of over one hundred and fifty miles. The Uncompahgre Valley is fertile, and along the branch of the Denver & Rio Grande Railroad from Montrose to Ouray, is under high state of cultivation. The cereals, fruit and vegetables, together with forage plants, flourish here in the greatest luxuriance. Here was the Indian reservation,

CHIPPETA FALLS IN THE BLACK CAÑON.

and here lived Ouray, the friend of the white man. It is only a few years since the good chief died, and his farm and buildings are still pointed out to the traveler, on the line to the town of Ouray, about two miles south of Montrose. The land in the valley surrounding Montrose is gradually being brought under cultivation. Irrigating canals have been constructed, and the rich soil responds generously to the demands of the farmer. Mining and pastoral industries also contribute greatly to the success of Montrose. There can be found excellent hunting and fishing in the vicinity.

A UTE COUNCIL FIRE.

Delta is twenty-one miles from Montrose, and is the county seat of Delta County. It is situated in the delta formed by the junction of the Uncompahgre and the Gunnison Rivers. The town is in a fine agricultural region and is supported by farming, pastoral and mining industries. It is destined to become, in time, a considerable business centre, (Population, 400. Distance from Denver 374 miles. Elevation, 4,980 feet.)

Between Delta and Grand Junction there are a number of small stations which will not interest the traveler, but the scenery through which the railroad passes (while it is not especially startling) will interest him. After passing Delta the road crosses the Uncompahgre and follows the west bank of the Gunnison (the same river that was left at Cimarron, forty-four miles behind us). In about five miles we cross to the east bank of the Gunnison and roll along beneath cliffs which tower on our right above the train, leaving but little room between rocks and river. At Bridgeport the cars plunge into the Bridgeport Tunnel, 2,256 feet in length, the longest tunnel on the Denver & Rio Grande Railroad. Shortly an iron bridge over a fine stream (the Gunnison River) is passed, and we find ourselves at the junction of the Gunnison with the Grand River.

> **GRAND JUNCTION.**
> Chief City of
> Grand River Valley,
> at
> Junction of Grand and
> Gunnison Rivers.
> Population, 1,500.
> Distance from Denver,
> 425 miles.
> Elevation, 4,594 feet.
> Eating Station.

In the Valley of the Grand River, and surrounded by a fertile and well watered country, Grand Junction is destined to become the leading city of western Colorado. An extensive system of irrigating ditches has been established, and all the land under these ditches taken up and most of it cultivated. The comparatively low altitude of this valley, it being the lowest among the Rocky Mountains with but one exception in Utah, makes it especially adapted to the cultivation of fruit. Peaches, grapes, apricots, pears and small fruits flourish here in great luxuriance, and most of the farmers have planted orchards and vineyards of greater or less extent. The usual farm products thrive in the valley, and large crops can be counted on with the greatest confidence. Grand Junction is the county seat of Mesa County, and has business and public buildings of a substantial character. Shade trees have been planted on each side of the streets, giving the town a most pleasing and attractive appearance. There is one thing sure about the Grand River Valley, and that is it will never want for water, and with plenty of water for irrigation secured, the future prosperity of the valley and the consequent growth of Grand Junction are both assured. Back in the hills great herds are pastured, and mining is, though to a moderate extent, tributary to the town.

Fruitvale is the next station to the west, and while the town does not appear to amount to a great deal, yet the experiment which is being carried on here is of interest to all. The post office is called Fruita, though the railroad station has been named Fruitvale. The post office and the station would have been given the single name of Fruitvale but for the fact that there are other "Fruitvale" post offices and the government does not care to multiply duplicate names The experiment carried on here, to which reference has been made, is that of fruit culture, the effort being to prove this valley as well fitted for this purpose as Utah. So far the experiments have been successful. (Population, 25. Distance from Denver, 436 miles. Elevation, 4,523 feet.)

The Colorado Desert. For a stretch of about two hundred and fifty miles beyond Fruitvale no agricultural country will be seen—over one hundred miles of this, in fact, is known as the Colorado Desert. But well informed people assert that all this desert needs to be made fertile is irrigation. Water can be got on this land from the Grand river, and perhaps before another decade has passed away the "Colorado Desert" will be ranked with that geographical myth of twenty years ago. "The Great American Desert."

The Book Cliffs. The intervening psace of one hundred miles between the Grand River and the Green would be monotonous were it not for the glimpses one obtains, to the left, of the snow-crowned San Rafael and Sierra La Sal mountains and the constant presence, to the right, of the multiform and varicolored Book

GRAND CAÑON OF THE COLORADO.

Cliffs. These Cliffs are the northern shore of what in former ages must have been a great inland sea, across whose basin the railroad runs. They vary in altitude from seven thousand to nine thousand feet and divide the waters of the Grand River from those of the White, extending two hundred miles from east to west. There are no stations of any importance between Grand Junction and Green River, the train pausing in transit only for water.

Green River. This is an eating station, on the west bank of the Green River, and on alighting from the cars the traveler is astonished at the elegance of the hotel and the beauty of its surroundings, situated, as it is, away out on the edge of the desert. A handsome lawn of shaven grass surrounds the hotel, ornamented with trees and shrubs. All the modern conveniences are to be found within, even to the latest style of electric light, and one of the best meals to be found on the entire journey is here set before the traveler. The hotel buildings are owned by the railroad company and no pains have been spared to make everything first class. Green River is a shipping point of considerable importance for stock. (Population, 25. Distance from Denver, 544 miles. Elevation, 4,069 feet.)

Grand Canon of the Colorado. From the bridge across Green River the traveler, can, if the day is clear, catch a glimpse of the rugged walls of the Grand Cañon of the Colorado, scarcely fifty miles to the southward.

Climbing the Wasatch Range. From Green River to Soldier Summit, a distance of ninety-nine miles, the grade is a constant ascent, the scenery growing wilder and more varied as the advance is made. The road extends to the northward, and, after passing Sphinx, Desert Switch and Cliff Siding, unimportant side tracks, reaches Lower Crossing, twenty-five miles from Green River.

Lower Crossing is situated on Price River in the midst of interesting scenery, stock raising is tributary to the town. (Population, 25. Distance from Denver, 570 miles. Elevation, 4,630 feet.)

Price. Situated on the south fork of the Price River, the town has a very fertile valley, though of limited extent surrounding it. What arable land there is has been carefully utilized, and large crops of potatoes, alfalfa, oats and vegetables are raised here, through the aid of irrigation. There are mines of asphaltum to the northward, which are worked extensively, and the product shipped to the east. Price is also an important shipping point for cattle and sheep. The scenery here is very attractive, and the hunting and fishing are excellent. (Population, 100. Distance from Denver, 611 miles. Elevation, 5,547 feet.)

Fort Dushane. Eighty miles to the northward from Price, on the Uintah and Uncompahgre Indian reservation, is Fort Dushane, the Government post, supplies for which are forwarded from Price. Fort Dushane, has four companies of infantry and two of cavalry, numbering in all three hundred men. There are 4,000,000 acres in the reservation, all of which are at the service of only 2,500 Indians.

CASTLE GATE,
Entrance to
Price River Cañon.
Height, 500 feet.

Six miles beyond Price station the train enters the famous portals of Castle Gate, which stand at the entrance of the Price River Cañon. Castle Gate is similar in many respects to the gateway in the Garden of the Gods. The two huge pillars, or ledges of rock composing it, are offshoots of the cliffs behind. They are of different heights, one measuring five hundred, and the other four hundred and fifty feet, from top to base. They are richly dyed with red, and the firs and pines growing about them, but reaching only to their lower strata, render this coloring more noticeable and beautiful. Between the two sharp promontories, which are separated only by a narrow space, the river and the railway both run, one pressing closely against the other. The stream leaps over a rocky bed, and its banks are lined with tangled brush. Once past the gate, and looking back, the bold headlands forming it have a new and more attractive beauty. They are higher and more massive, it seems, than when we were in their shadow. No other pinnacles approach them in size or majesty. They are landmarks up and down the cañon, their lofty tops catching the eye before their bases are discovered. It was down Price River Cañon, and past Castle Gate, that Sidney Johnston marched his army home from Utah. For miles now, and until the mountains are crossed, the route chosen by the General is closely followed. The gateway is hardly lost to view by a turn in the cañon before we are scaling the wooded heights. The river is never lost sight of. The cliffs which hem us in are filled with curious forms. Now there is seen a mighty castle, with moats and towers, loopholes and wall; now a gigantic head appears. At times side cañons, smaller than the one we are in, lead to verdant heights beyond, where game of every variety abounds.

CASTLE GATE.

Kyune. Distance from Denver, 632 miles. Large stone quarries are worked here.

Pleasant Valley Junction. This little town is situated in the midst of rich and extensive coal measures. A branch road runs to the coal mines a distance of about twenty miles to the southward. The coal is valuable for coking, and is used in the various smelters of the territory. (Population, 200. Distance from Denver, 636 miles. Elevation, 7,177 feet.)

Coal Branch. From Pleasant Valley Junction the Coal Branch extends to Mud Creek, a distance of 20 miles. The intervening stations are Hale, Schofield and Coal Mine. The chief business of the road is the transportation of coal, which is mined extensively here.

Soldier Summit. Here we are on the highest railroad point on the Wasatch Range. Good pasturage covers the mountain tops, and great herds of cattle, horses and sheep graze here among the sage brush. The scenery here is wild and picturesque and the view is wide, embracing a great sweep of serrated mountain summits. (Population, nominal. Distance from Denver, 642 miles. Elevation, 7,465 feet.) From this point the descent is made to the Utah Valley.

Red Narrows. Here the cliffs rise on each side of the track, assuming fantastic forms, and glowing with varied colors, among which red is predominant; hence, the name.

Spanish Fork Canon is charmingly picturesque, and a spot which would delight the artist. It is characterized by fresh foliage, soft contours, charming contrasts, and sparkling waters Emerging from the cañon the traveler realizes that one stage of his mountain journey has been achieved, and before him lies one of the most fertile valleys in the world.

Utah Valley. This favored spot presents the appearance of a well-cultivated park. It has an Arcadian beauty, and resembles the vales of Scotland. In its centre rests Utah Lake, where

" . . . the stars an d mountains view
The stillness of their aspect in each trace
Its clear depth yields of their far height and hue."

A little back from the lake stand the towns of Provo and Spsingville, shaded by the near peaks of the range. Utah Valley possesses a fertile soil, a delightful climate, and is one of the best farming sections of Utah. Fruit trees and grape vines grow as readily as hay and cereals. Eastward the oblong-shaped basin is shut in by the Wasatch Mountains; and on the west is the Oquirrh Range. Northward are low hills, or mesas, crossing the valley and separating it from that of the Great Salt Lake; while in the south, the east and west ranges approach each other and form blue-tinted walls of uneven shape. To the left of this barrier Mount Nebo, highest and grandest of the Utah peaks, rises majestically above all surroundings. Its summit sparkles with snow, its lower slopes are wooded and soft, while from it, and extending north and south, run vast, broken, vari colored confreres. The valley is like a well-kept garden; farm joins farm; crystal streams water it; and scattered about in rich profusion are long lines of fruit trees, amid which are trim, white houses. All these evidences of prosperity testify to the virtues of industry, frugality and perseverance, which no one can deny are possessed by the Mormon farmers.

Spanish Fork. This is the first town in Utah Valley that the west-bound tourist enters. It is situated on the Spanish Fork River, and is a most pleasant rural village. Fruit and shade trees abound. Agricultural, horticultural, and pastoral industries are pursued by the inhabitants. Vineyards flourish, wine is made, dairy products are a specialty, and the cereals and all kinds of vegetables are cultivated. (Population, 2,500. Distance from Denver, 679 miles. Elevation, 4,721 feet.)

Springville. This is another typical Mormon town. It is only four miles from Spanish Fork, and naturally possesses similar characteristics. The town derives its name from the fact that a strong hot spring pours its waters into a stream just above the town, in Hobble Cañon. The water does not freeze in winter, and thus a flouring mill run by it is enabled to work the year 'round. (Population, 2,500. Distance from Denver, 683 miles. Elevation, 4,565 feet.)

A QUIET NOOK

PROVO,

County Seat of Utah Co.

Summer Resort.

Population, 5,000.

Distance from Denver, 689 miles.

Elevation, 4,517 feet.

This pretty little city belongs to the best type of Mormon towns, and a description of it will serve to give the reader a good idea of the characteristics of all the towns built by the Mormons. The dwellings, as a rule, are comfortable, but not imposing in appearance. Many of them are constructed of *adobe* or sun-dried bricks, and all are situated in lots of generous proportions and surrounded by ornamental and fruit trees. Water for irrigating purposes flows down each side of the streets, and shade trees in abundance and of luxuriant growth render the walks cool and inviting. Gardens filled with fruits, flowers and vegetables are the rule, and a quiet, peaceful, industrious semi-rural life is the good fortune of the residents here. The town is eminently fitted for a health and pleasure resort, and has also great advantages as a manufacturing

SPANISH FORK CAÑON.

centre. The Timpanogas River furnishes unexcelled water power, while inexhaustible supplies of artesian water are to be found at a depth of from forty to two hundred feet. The city has, in fact, the finest water supply of any in Utah Territory. Provo has a fine public school system and is the seat of the Brigham Young Academy, which was amply endowed by the first President of the Mormon Church, from whom the school takes its name. Its churches and public buildings, including an opera house, are a credit to its people, who are of a literary taste and inclined to liberality of thought. Utah Lake, a fine body of fresh water, lies to the southwest, and to the north and east are the Wasatch Mountains. Farming, horticulture and the raising of cattle and sheep are tributary industries, while in the town are large saw mills, flouring mills and woolen mills, the most extensive in Utah.

Utah Lake. Mention has already been made of this beautiful body of water, but the statistical traveler may want to know something more definite about its dimensions. The lake is thirty miles long, six miles wide, and is fed by the

TRAMWAY IN LITTLE COTTONWOOD CAÑON

American Fork, Spanish Fork and Provo Rivers, and Salt, Peteetweet and Hobble Creeks. Its outlet is the Jordan River which, flowing northward, empties into Great Salt Lake. There are plenty of fish in Utah Lake, chiefly trout and mullet.

American Fork. On the western extremity of Utah Lake, is American Fork, a thriving town beautifully situated and embowered in trees. Agricultural and pastoral industries are tributary to its prosperity. (Population 1,800. Distance from Denver, 702 miles. Elevation, 4,567 feet.)

Lehigh. Three miles from American Fork is Lehigh, another thriving town also on Utah Lake. Fruit and shade trees abound and make the town a place of sylvan beauty. The same industries thrive here as in the sister town mentioned above. (Population, 2,000. Distance from Denver, 705 miles. Elevation, 4,544 feet.)

Bingham Junction. This station is at the junction of the Bingham and Alta branches of the road and, therefore, is quite a bustling place in the way of railroad business, though it has but a nominal population. (Distance from Denver, 723 miles. Elevation, 4,366 feet.)

Bingham Branch. This branch extends southwest to Bingham, a distance of sixteen miles. The intervening stations are Revere, Lead Mine and Terra Cotta.

Bingham. The town may almost be classed as a suburb of Salt Lake City, as it is less than an hour's ride from the capital of Utah Territory. The main industry of the surrounding population is mining. (Population, 900. Distance from Denver, 740 miles. Elevation, 4,375 feet.)

Alta Branch. This branch extends to the northward from Bingham Junction to Alta, a distance of thirty-five miles. The intermediate stations are Sandy and Wasatch. The line passes through the Little Cottonwood Cañon en route.

Alta. This is a mining town known all round the world. The place is not only entertaining in itself, but in its neighborhood are a large number of easily accessible gorges, lakes and hilltops full of artistic material and of trout fishing; or, if the tourist goes late in the season, of good shooting and ample opportunity for dangerous adventures in mountaineering. The Little Cottonwood cañon is one of those great crevices between the peaks of the Wasatch Range, plainly visible from Salt Lake City, and distinguished by its white walls, which, when wet with the morning dews, gleam like monstrous mirrors as the sunlight reaches them from over the top of the range.

The River Jordan. After the valley of Utah Lake has been left behind, en route to Salt Lake City, on the left of the track is seen a small river of yellow water meandering through the sage brush and volcanic scoria. The river is the Jordan, so called because it connects the Utah with the Great Salt Lake, as its namesake does Galilee and the Dead Sea.

SALT LAKE CITY,
Capital of Utah Territory.
Population, 25,000.
Elevation, 4,228 feet.
Distance from Denver, 735 miles.

Forty-one years ago Brigham Young stood on Ensign Peak, the "Mount of Prophesy," and announced to his followers that down in the valley below should be founded the new "City of Zion," the future home of the Latter Day Saints. Up to 1871 the original settlers virtually lived apart from the rest of the world. This was owing to the religious views of the Mormons, which made them a peculiar and isolated people. To mining is due the first incursion of Gentile population, which population has steadily increased, until at present the community of Salt Lake City differs but little from any other in its social, business or religious aspect, except that it possesses, in addition to the accepted religious associations which exist elsewhere, one which differs from all others. The city is situated at the base of the Wasatch Mountains, which are a part of the great Continental Range dividing the Far West from the plains which extend from the base of the Rockies to the Missouri River. The finest residence portion of the city occupies the mountain bench, once the shore of a great inland sea, from which, ages ago, the waters receded until they settled in the basin of the Great Salt Lake, distant eighteen miles from the water marks yet plainly to be seen above the city. The location is such as to command a view of the entire valley, both ranges of mountains, and the southern portion of the lake. The streets are one hundred and thirty-

BIRD'S-EYE VIEW OF SALT LAKE CITY, UTAH.

two feet wide and bordered on each side with long rows of shade trees. Streams of pure water are conducted in ditches along both sides of all the streets. The business sections are well built. One of the largest business enterprises of the city is the Coöperative Establishment. For convenience it is universally called the "Co-op," its title in full is the "Zion's Coöperative Mercantile Institution." It has a central building for headquarters and branches throughout the city and Tertority Whenever one sees a building with the mystic initials "Z C. M. I." on its sign, one may know it is a branch of the great "Co-op." The headquarters of this institution are of brick, three hundred and eighteen by fifty-three feet in size, three stories high, and built over a large cellar. This building is crowded with merchandise of every description, and does an extensive wholesale and retail business. "Temple Square" is a great attraction for the tourist. Here are situated the Mormon Temple, Tabernacle and Assembly Hall. The Tabernacle is immense in its proportions, the roof resembling an upturned boat, and is visible from nearly every part of the city. The Temple is still unfinished, but even now its massive walls of granite bespeak the future magnificence of the edifice. Near by is the Bee Hive, once the home of Brigham Young and opposite the house of President Taylor. The **Hot Springs** of Salt Lake are highly medicinal, and the large baths are resorted to for many ailments. Within a short radius of the city the attractions are varied and numerous. Fort Douglass, the Lake, Emigration City, Bingham, Little and Big Cottonwood Cañons are easily reached. From Ensign Peak a panoramic view of the surrounding country is had. One may look from it down the greater part of Utah's length, while near at hand lie the city and lake. The Fort is also a popular resort, and not only commands an extensive view, but affords excellent opportunities of studying garrison life. The rides, drives and rambles are innumerable. Every taste is catered to. For those who love grandeur, there are the mountains, with their narrow trails, secluded parks, wild cañons and deep gorges; for those preferring gentler aspects, the valley, glowing with freshness, affords continual pleasure; for those craving the mysterious, there is the lake, large, silent and strange. The hotels are excellent, the climate unexcelled, and days may be passed delightfully in exploring and in studying the wealth of attractions. There are theatres, reading rooms, good horses, perfect order and universal cleanliness. Many of the private houses are palatial, and altogether the city is one of rare beauty and interest.

GREAT SALT LAKE.
Area, 2,500 square miles.
Mean Depth, 20 feet.
Specific Gravity, 1.107.
Length, 126 miles.
Breadth, 45 miles.

As far as can be learned, the first mention in history of the Great Salt Lake was by the Baron La Houtan, in 1689, who gathered from the Western Indians some vague notions of its existence. Capt. Bonneville sent a party from Green River in 1833 to make its circuit, but they seem to have given up the enterprise on reaching the desert on the northwest, on which they lost their way, and after weeks of aimless wandering found themselves in Lower California. To General John C. Fremont must be given the credit of first navigating its waters. In 1842, on his way to Oregon, General Fremont pushed out from the mouth of Webber River, in a rubber boat, for the nearest island. He found it to be a desolate rock, fourteen miles in circumference and named it Disappointment Island. Captain Stansbury, on a subsequent visit, renamed it Fremont's Island, which name is retained. In 1850 Captain Stansbury spent three months in making a detailed survey of the Lake, its shores and islands. In brief he

THE GREAT SALT LAKE.

found the west shore a salt-encrusted desert; the north shore composed of wide salt marshes, overflowed under steady winds from the south; the east shore possessed good, irrigable lands; the south shore was set with mountain ranges standing endways towards the lake, with the grassy valleys, Spring, Toelle and Jordan, intervening. The principal islands are Antelope and Stansbury. rocky ridges ranging north and south, rising abruptly from the water to a height of three thousand feet. Antelope is the nearest to Salt Lake City and is sixteen miles long. Stansbury is twenty miles to the westward and is twelve miles in length. Both have springs of fresh water and good range for the stock, with which they are now covered. Of minor islands there are Fremont, Carrington, Gunnison, Dolphin, Mud, Egg and Hat, besides several small insular promontories without names. The first white man's boat to navigate the lake was probab'y that of Fremont; Captain Stansbury came next with his exploring boat curiously named the "Salicornia"; next in order were the Walker brothers, merchants of Salt Lake City, who sailed for some years a lonesome pleasure yacht. There is now a considerable yachting fleet, which is yearly growing in size. The lake covers an area of 2,500 square miles. Its mean depth does not probably exceed twenty feet, while the deepest place between Antelope and Stansbury is 60 feet. These two principal islands used to be accessible from the shore by wagon, but now boats must be used. From 1847 to 1856 the lake gradually filled five or six feet and then slowly subsided to its old level. In 1863 it began to fill again and in four or five years reached a point considerably higher than its present level, perhaps four or five feet. In the year 1875 a pillar was set up at Black Rock, by which to measure the rise and fall, resembling a tide, but having no ascertained time. It is very slight compared with what it formerly was. Professor Gilbert, of the Geological Survey, says that twice within recent geological time it has risen nearly a thousand feet higher than its present stage, and, of course, covered vastly more ground. He calls that lake after Captain Bonneville, the original explorer of these regions and whom Irving has immortalized, Lake Bonneville. Causes which learned men assign as producing what they call a glacial period might easily fill the lake until it extended nearly the whole length of Utah. During the last high stage, Professor Gilbert says there were active volcanoes in it. It is generally agreed that its first outbreak was via Marsh Creek, and the Portneuf into the Snake At the present height of that channel (where the Utah and Northern passes out of Cache Valley) it remained a long time stationary and then seems to have receded rapidly to a second stationary point, and so on down to its present stage. There is one very heavy beach-mark on all the hills surrounding its extended area and on the hills, which were then islands, and a curious thing is the fact that this beach-mark varies in altitude from one hundred to three hundred feet, showing that the earth in this valley is still far from having reached a stable equilibrium.

The most mysterious thing about this inland sea, aside from its saltness, is the fact that it has no known outlet. A great number of fresh water streams pour into the lake from all sides, yet the water remains salt and the lake does not overflow. The saline or solid matter held in solution by the water varies as the lake rises and subsides. In 1842 Fremont obtained "fourteen pints of very white salt" from five gallons of the water evaporated over a camp fire. The salt was also very pure, assaying 97.80 fine. In 1850 Dr. L. D. Gale analyzed a sample of it which yielded 20 per cent. of pure common salt, and about 2 per cent. of foreign salts, chlorides of lime and magnesia. Sergeant Smart, U. S. A., analyzed a sample in 1877, and found an imperial gallon to contain nearly 24½ ounces of saline matter, amounting to 14 per cent., as follows :

ASSEMBLY HALL, TABERNACLE AND TEMPLE, SALT LAKE CITY.

Common salt	11.735
Lime carbonate	.016
Lime sulphate	.073
Epsom salt	1.123
Chloride of magnesia	.843
Percentage of solids	13.790
Water	86.210
	100.

One hundred grains of the dry solid matter contained:

Common salt	85.089
Lime carbonate	.117
Lime sulphate	.531
Epsom salt	8.145
Chloride of magnesia	6.118
	100.

It compares with other saline waters about as follows:

	Water.	Solids.
Atlantic Ocean	96.5	3.5
Mediterranean	96.2	3.8
Dead Sea	76.	24.
Great Salt Lake	86.	14.

And in specific gravity, distilled water being unity:

Ocean water	1.026
Dead Sea	1.116
Great Salt Lake	1.107

The solid matter in the water varies between spring and fall, between dry and wet seasons, and also between different parts of the lake, for nearly all the fresh water is received from the Wasatch on the east. It is the opinion of salt makers that an average of the lake at its present stage would show the presence of 17 per cent of solid matter.

Within a comparatively recent date, Salt Lake has become a fashionable bathing resort. In the long sunny days of July, August and September, the water becomes deliciously warm, much warmer in fact than the ocean, and this pleasant temperature is reached a month earlier and remains a month later. The water is so dense that one is sustained without effort, and vigorous constitutions experience no inconvenience from remaining in it a long time. A more delightful

BEE HIVE HOUSE.

and healthy exercise than buffeting its waves when it is a little rough can hardly be imagined. There are two popular bathing resorts on the Lake, near Salt Lake City.

Lake Park is situated on the Denver & Rio Grande Railroad, seventeen miles from Salt Lake City, and nineteen from Ogden. Located as it is between the two most important towns in the territory, Lake Park is in a position to command a large patronage from both cities. During the season bathing trains are run almost hourly from Salt Lake City to the Park, these trains make it possible for all transcontinental travelers stopping off at Salt Lake City to have a bath in the great dead sea. Each of the elegant bath rooms is fitted with fresh water shower bath, stationary wash bowls, mirrors, chairs, incandescent electric lights, etc., making Lake Park one of the most attractive watering places on the continent. There is a first class restaurant and exchange, and in the elegant Moorish pavillion on the lake shore, a band plays popular music to which the visitors can dance on the wide and level floor entirely free of extra expense. All through trains stop at this charming resort.

Garfield Beach is the old bathing resort, twenty miles west of Salt Lake City. It is situated on the Utah & Nevada Railroad, and great improvements have recently been made here. There is a large and commodious hotel, and extensive bath houses have been erected. Garfield Beach shares with Lake Park in the esteem and patronage of the people.

Salt Lake to Ogden. From Salt Lake to Ogden the Denver & Rio Grande Railroad traverses a narrow plain. On the west lies the Great Salt Lake, while to the north rise the serrated peaks of the Wasatch Mountains. This region is under a high state of cultivation. Farms reach their golden or green fields over its length and breadth, and little streams run in bright threads out of the mountain cañons down across the meadows. The lake is in full view of the traveler most of the way, and is a never-ending source of interest. The train speeds on, and entering an amphitheatre, set around with mountains, reaches Ogden, the western terminus of the Denver & Rio Grande and Union Pacific Railroads. (Population, 10,000. Distance from Denver, 771 miles. Elevation, 4,286 feet.)

QUEEN S CAÑON

GRAND CAÑON, FROM TO-RO WASP.

PUEBLO TO ALAMOSA.

FROM Pueblo to Cuchara Junction, a distance of 75 miles, the railroad extends to the southward across the plains, which stretch in one vast unbroken expanse to the eastern horizon, while to the west lies the Greenhorn Range with its intervening foothills.

Spanish Peaks. To the south rise the famed Spanish Peaks, springing directly from the plains, remarkable for their symmetry of outline, and reaching an altitude respectively of 13,620 and 12,720 feet. The Indians, with a touch of instinctive poetry, named these beautiful mountains "Wahatoya," or twin breasts. As a matter of orthographical interest, the reader may be pleased to know that the Indian spelling of the word is as follows: "Huacjatollas!"

Trinidad Branch. From Cuchara Junction, one line of the road extends in a southern direction to Trinidad, the largest city in Southern Colorado and the centre of the famous coal measures of El Moro.

This branch of the road does not pass directly through grand scenery, as it extends to the southward across the plains, and to the east of the mountains ; but the line is of great commercial importance, as by its connections at Trinidad it affords a direct through route to the Gulf of Mexico. Locally, also, it is of especial importance as El Moro and Trinidad are in the heart of one of the greatest coal regions in the west, and the agricultural and pastoral industries of the plains are of large proportions. From Cuchara Junction the stations occur in the following order: Tuna, Rouse Junction, Santa Clara, Boaz, Apishapa, Barnes, Chicosa and El Moro.

El Moro is worthy of special mention because of its extensive coal mines and coking ovens; the latter are 250 in number, and the greatest in the State. The town derives its name from the great butte (El Moro) which towers above it, presenting a very striking object to the view. (Population, 200. Distance from Denver, 206 miles. Elevation, 5,879 feet.)

TRINIDAD.
Commercial and Manufacturing City.
Population, 6,000.
Elevation, 5,994 feet
Distance from Denver, 210 miles.

This is the metropolis of southeastern Colorado, and the terminus of this branch of the Denver & Rio Grande Railroad. Trinidad is the trade and money centre for an immense territory, including portions of northern Texas, southern Colorado and northern New Mexico. In natural resources, Trinidad is exceedingly rich, being the centre of the largest coal belt in the world, and the supply depot for most of the coke used in the Great West. In addition to coal and coke in the immediate vicinity, iron exists in unlimited quantities. The supply of gypsum, granite, alum, fire-clay, silica, grit or grindstone, limestone and the finest of building stone is absolutely inexhaustible. Trinidad, from the natural deposits alone, must of necessity become a manufacturing centre of vast importance, and has already taken advance

VETA PASS AND DUMP MOUNTAIN.

steps in this regard. A $200,000 rolling mill is now under way. The manufacture of cement, mineral paint, lime, and plaster of paris, are all important industries, while the production of building brick is very large in its proportions. Fire-brick and silica brick will soon be an additional industry. In and around Trinidad no less than three thousand laborers are now employed, and this large and daily increasing number of men spend their money in Trinidad. The city has water-works, gas-works, electric light, street cars, and other metropolitan improvements. The schools and churches are very superior, while the business houses and residences are a credit to the city. Its elevation above the level of the sea insures a delightful climate, free from malaria and other poisons common to lower altitudes, while the scenic surroundings are unsurpassed. Raton Peak and the distant range adding their grandeur to the beauty of the scene. Trinidad is a railroad centre, with three great trunk lines already in operation, with three more moving toward it; is the most important wool centre in Colorado, being the original market for 3,000,000 pounds, and is also a great cattle centre and, for that reason, the largest hide and pelt-receiving point of the State. Resuming the journey to Alamosa, the tourist returns to

Cuchara Junction. A small town at the junction of the New Mexico and Trinidad extensions of the Denver & Rio Grande Railroad. The supporting industries are pastoral and agricultural pursuits. (Population, 25. Distance from Denver, 169 miles. Elevation, 5,942 feet.)

Walsens. A flourishing town doing a large business, both at home and abroad. It is surrounded by a fine pastoral country, and also derives revenue from agriculture. Coal is mined near here in large quantities. (Population, 1,000. Distance from Denver, 176 miles. Elevation, 6,189 feet.)

La Veta. A prosperous village, surrounded by a pastoral country and in the midst of most beautiful scenery, being near the foothills of La Veta Mountain and the famous pass known by the same name. The Spanish Peaks are also in plain view to the east. (Population, 300. Distance from Denver, 191 miles. Elevation, 7,024 feet.)

VETA PASS.

Elevation, 9,393 feet.

Maximum Grade, 211 feet to the mile.

Distance Across Pass, 13 miles.

The ascent of this famous pass is one of the great engineering achievements of the Denver & Rio Grande Railroad. The line follows the ravine formed by a little stream, La Veta Mountain rising to the right. At the head of this gulch is the wonderful "Mule-Shoe Curve," the sharpest curve of the kind known in railroad engineering. In the centre of the bend is a bridge, and the sparkling waters of the mountain stream can be seen flashing and foaming in their rocky bed below. Standing on the rear platform of the Pullman car as the train rounds the curve, the tourist can see the fireman and engineer attending to their duties. From this point the ascent of Dump Mountain begins, rocks and precipitous escarpments of shaley soil to the right and perpendicular cliffs and chasms to the left. The ascent is slowly made, two great Mogul engines urging their iron sinews to the giant task. The view to the eastward is one of great extent and magnificence. The plains stretch onward to the dim horizon line like a gently undulating ocean; from which rise the twin cones of Wahatoya, strangely fascinating in their symmetrical beauty. At the summit of the pass the railroad reaches an elevation of 9,393 feet above the sea.

Veta Mountain is to the right, as the ascent of the pass is made, and rises

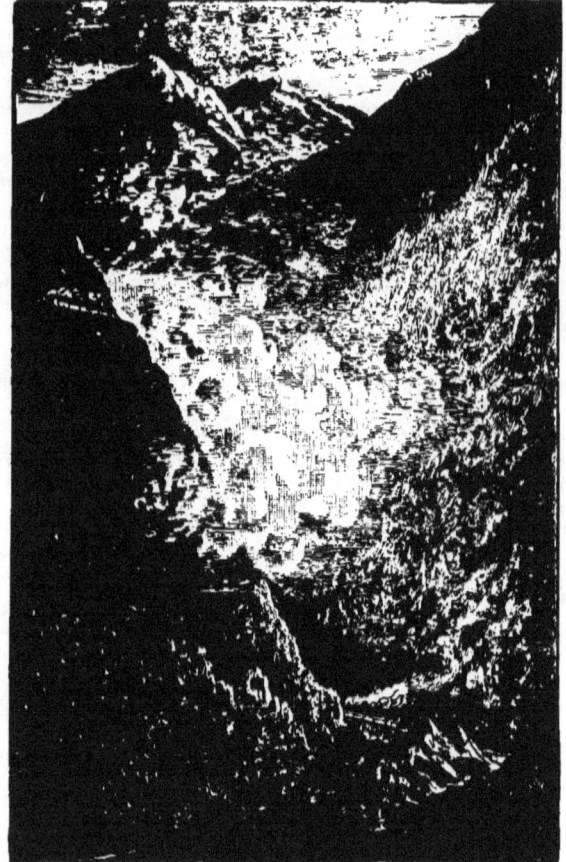

CLIMBING THE MOUNTAINS AT VETA PASS.

with smooth sides and splintered pinnacles to a height of 11,176 feet above the sea. The stupendous proportions of this mountain, the illimitable expanse of the plains, the symmetrical cones of the Spanish Peaks present a picture upon which it is a never-ceasing delight for the eye to dwell. The train rolls steadily forward on its winding course, and at last reaching the apex, glides into the timber and halts at the handsome stone station over 9,000 feet above the level of the distant sea. The downward journey is past Sierra Blanca and old Fort Garland and through that pastoral and picturesque valley known as San Luis Park.

Placer. At Placer, one can say that the descent of Veta Pass has been accomplished, although it is still down grade as far as Alamosa. This little town is situated on the eastern border of the San Luis Valley and at the western extremity of La Veta Pass. Good hunting and fishing can be found in the neighboring foothills. The tributary industries are agriculture and stock raising. (Population, 75. Distance from Denver, 212 miles. Elevation, 8,410 feet.)

Garland. This town was formerly known as Fort Garland, and was a United States military post. Sierra Blanca, elevation, 14,464 feet, the highest mountain in the United States with one exception, is seventeen miles distant. Good trout fishing and shooting can be found in the adjacent foothills. Garland's tributary industries are agriculture and stock raising. (Population, 200. Distance from Denver, 325 miles. Elevation, 8,945 feet.)

> SIERRA BLANCA
> Highest Mountain
> of
> The Rocky Range.
> Elevation,
> 14,464 feet,

Sierra Blanca is the monarch of the Rocky Range, and is characterized by the peculiarity of a triple peak. The mountain rises directly from the plain to the stupendous height of 14,464 feet, over two miles and three-fifths of sheer ascent. A magnificent view of this mountain is obtained from the cars as soon as the descent from Veta Pass into the San Luis Valley has been made. Surely it is worth a journey across the continent to obtain a view of such a mountain! Although a part of the range, it stands at the head of the valley, like a monarch taking precedence of a lordly retinue. Two-thirds of its height is above timber-line, bare and desolate, and except for a month or two of midsummer, dazzling white with snow, while in its abysmal gorges it holds eternal reservoirs of ice.

> "Oh, sacred mount with kingly crest
>   Through tideless ether reaching,
> The earth world kneels to hear the prayer
>   Thy dusky slopes are teaching.
> With mystic glow on sunset eyes
> All trembling lie thy blood-red leaves,
> Their silken veins with gold inwrought.
> Oh, glorious is thy world-wide thought!"

The lower slopes of the mountain are clad in vast forests of pine and hemlock, while its grand triad of gray granite peaks lift into the sky their sharp pyramidal pinnacles, splintered and furrowed by the storm-compelling and omnipotent hand of the Almighty. To the north and south, for a distance of nearly two hundred miles, it is flanked by the serrated crests of the Sangre de Cristo Range, the whole forming a panorama of unexampled grandeur and beauty.

San Luis Park. This great and fertile valley is located in Southern Colorado, bordering New Mexico, and is drained by the Rio Grande, one of the largest of Colorado's rivers, into which flows from the lofty mountain ranges surrounding the park, almost numberless little mountain streams. This park, which was once the bottom of a vast mountain lake, contains fully 10,000 square miles — equal to the entire area of Massachusetts. The soil is alluvial, from six to fifteen feet deep, and the surface is naturally well adapted for irrigation, which the rivers and streams in the park are abundantly capable of providing. The park, or valley, as it is frequently called, is from 7,000 to 7,300 feet above sea level. This elevation insures a light, pure atmosphere, free from all malarial conditions, and especially favorable for those disposed to pulmonary affections. The climate is cool in the summer, and not severe in the winter — scarcely ever more than an occasional snowfall of two or three inches in the valley. Too much in praise of the attractions and beauty of the climate of the San Luis Valley cannot be said. The grand chains of mountains, which entirely surround the park, present scenery unsurpassed in the world. Spring wheat will yield from thirty to fifty bushels to the acre, oats from fifty to seventy-five bushels, peas from thirty to forty bushels, potatoes from

SIERRA BLANCA—THE HIGHEST MOUNTAIN IN COLORADO.

two hundred to three hundred bushels to the acre ; beans, cabbage, all kinds of root crops, are unexcelled anywhere. Hops do well ; tomatoes and melons are grown, but with some effort. Corn, in consequence of the elevation, except for garden purposes, does not pay. Alfalfa—the clover of the mountains—does well, yielding from four to six tons in two cuttings. Common red clover, timothy and red top, do well. The native grasses, by irrigation, yield two tons per acre. All kinds of small fruit do exceedingly well. Grapes are untried, but it is believed they will succeed. Apples and cherries do well; plums and pears may, but peaches cannot, be grown. Surrounding the valley, embracing the foot hills and lower mountain ranges, is a range covering millions of acres, where cattle, horses and sheep can feed for nine months in the year. The grasses are more abundant and nutritious than upon the lower elevations. The stock so grazed upon these free ranges in the summer and fed upon the home farms in the valley in the winter, can be handled without hazard, and with certainty of profitable return to the farmer and large ranchmen.

ALAMOSA.
Junctional City.
Eating Station.
Population, 1,200.
Distance from Denver, 250 Miles.
Elevation, 7,546 feet.

This is one of the most considerable towns of the San Luis Valley. It is situated on the west bank of the Rio Grande River and at the junction of the New Mexico & Wagon Wheel Gap branches of the Denver & Rio Grande Railroad. The resources of the San Luis Valley have been described above, and it goes without saying that these resources are naturally tributary to the welfare of Alamosa.

The town is well supplied with stores of all kinds, some of which carry large stocks of goods. Great quantities of hay and grain, and farm produce generally, are shipped from this station, which also commands a large local trade. Within a short distance of the town a natural gas supply has been discovered, which only needs adequate development to make it an element of great prosperity to the city. The eating house at Alamosa, while unpretentious in its exterior, furnishes one of the best meals to be obtained anywhere, and has a wide-spread and well-deserved reputation. The scenery surrounding the town is grand, and the near proximity of the river makes it a favorite resort for sportsmen.

Wagon Wheel Gap Branch. From Alamosa a branch of the Denver & Rio Grande extends up the valley a distance of sixty-one miles to the great hot springs at Wagon Wheel Gap. The line passes through an exceedingly fertile agricultural country lying on both sides of the Rio Grande and irrigated by the great canals taken out from the river. In the proper season of the year thousands of acres of wheat and oats, alfalfa and other farm produce can be seen growing in the greatest luxuriance on both sides of the track.

Monte Vista. This flourishing town is an example of rapid growth and a proof of the self-sustaining character of the country. It is not yet five years old and is already beginning to assume the airs of a city. The surrounding country is full of coal, oil and gas. Very rich mines are being developed (ore running from $1,000 to $2,000 per ton) in the mountains southwest of Monte Vista, which is located in the midst of 300,000 acres of the richest irrigable land with abundance of water to supply it. Monte Vista is a new, growing, enterprising prohibition town and has a superior class of citizens. It is rapidly becoming an extra desirable residence locality. It has a first class roller process flouring mill, fifteen stores, two banks, a planing mill, three lumber yards, three weekly papers, three livery

SUMMIT OF VETA MOUNTAIN.

stables, large public library, an $8,000 school-house, a $75,000 hotel, seven church organizations, a secular Sunday society, secret societies, military company, cornet band, etc. In the vicinity is one farm of 7,000 and another of 4,000 acres. (Population, 1,000. Distance from Denver, 267 miles Elevation, 7,665 feet.)

Del Norte. This is the oldest town in what is known as the San Juan country and is the county seat of Rio Grande county. The town site was surveyed in 1872, though the town company was formed in 1871. The town is so situated as to be upon the line between the agricultural and mining sections. To the north and east of the town are the rich and rapidly settling agricultural and pastoral lands of the San Luis Valley; to the south and west are the great mines of San Juan. Del Norte is beautifully situated in a basin at the foot of the mountains, sheltered from the blasts of winter and having the most delightful weather in summer. The Rio Grande flows through the edge of the Del Norte town site, and offers to manufacturing interests exceptionally fine water power. Del Norte has some excellent business and dwelling houses, a fine public school building, two good church buildings—above the average, the Presbyterian College of the South-

WAGON WHEEL GAP.

west (a staunch educational institution), a fine flouring mill of the latest roller process, a large brewery using home-grown barley, two banks, a court house costing $30,000, the United States land office, where all business regarding lands in this district must be transacted, and countless other enterprises that cannot be mentioned here. On Lookout Mountain, 600 feet above the town, is mounted a large telescope, to be used in connection with the Presbyterian College of the Southwest. The view from the Lookout observatory is grand in the extreme. The streets of Del Norte are wide, and the town is noted for its growth of trees—mostly cottonwoods. Water for irrigating purposes is supplied by means of a main canal from the Rio Grande, with laterals over the town site along the sides of streets. The distance from Del Norte to the following points is: To Alamosa, 30 miles; to Saguache, 35 miles; to Villa Grove, 45 miles; to Monte Vista, 15 miles; to Veteran, 18 miles; to Summitville, 27 miles; to Wagon Wheel Gap, 30 miles; to Shaw's Springs, 6 miles; to Carnero, 25 miles. Del Norte is certainly a very attractive town. (Population, 1,200. Distance from Denver, 281 miles. Elevation, 7,880 feet.)

From Del Norte the line follows closely up the river amidst most attractive scenery. South Fork is a small station on the river and is a favorite stopping place for anglers.

WAGON WHEEL GAP HOT SPRINGS.
Distance from Denver, 311 miles.
Elevation, 8,449 feet.

The hot springs at Wagon Wheel Gap, together with the magnificence of the scenery, make it one of the most attractive pleasure resorts in Colorado. As the Gap is approached the valley narrows until the river is hemmed in between massive walls of solid rock, that rise to such a height on either side as to throw the passage into a twilight shadow. The river rushes roaring down over gleaming gravel or precipitous ledges. Progressing, the scene becomes wilder and more romantic, until at last the waters of the Rio Grande pour through a cleft in the rocks just wide enough to allow the construction of a road at the river's edge. On the right, as one enters, tower cliffs to a tremendous height, suggestive in their appearance of the palisades on the Hudson. On the left rises the round shoulder of a massive mountain. The vast wall is unbroken for more than half a mile, its crest presenting an almost unserrated sky line. Once through the gap the traveler looking to the south sees a valley encroached upon and surrounded by hills. Here is the old stage station, a primitive and picturesque structure of hewn logs and adobe, one story in height, facing the south, and made cool and inviting by wide-roofed verandas extending along its entire front. Not a hundred feet away rolls the Rio Grande, swarming with trout. A drive of

UP THE RIO GRANDE

TROUT FISHING AT WAGON WHEEL GAP.

a mile along a winding road, each turn in which reveals new scenic beauties, brings the tourist to the famous springs. The medicinal qualities of the waters, both of the cold and hot springs, have been thoroughly tested and proved to be of a very superior quality. Lieutenant Wheeler, U. S. A., gives the following analysis of these springs: No. 1 has a temperature of about 150° Fahrenheit, is bubbling continually, and is about eight feet wide by twelve feet long; No. 2 is a small bubbling spring, cold, about one foot in diameter, and gives out a strong odor of sulphuretted hydrogen; No. 3 is situated some distance from Nos. 1 and 2, at the foot of a hill; it bubbles continually and is of a temperature of 140° Fahrenheit. This spring is about three feet wide and the same in length; it is called the Soda Spring. In one thousand parts of the water of the springs of Wagon Wheel Gap are contained parts as follows:

	No. 1.	No. 2.	No. 3.
Sodium Carbonate	69.42	Trace.	144.50
Lithium Carbonate	Trace	Trace.	Trace.
Calcium Carbonate	14.08	31.00	22.42
Magnesium Carbonate	10.91	5.10	22.42
Potassium Sulphate	Trace.	Trace.	Trace.
Sodium Sulphate	23.73	10.50	13.76
Sodium Chloride	29.25	11.72	33.34
Silicic Acid	5.73	1.97	4.72
Organic Matter	Trace.	Trace.	
Sulphuretted Hydrogen	Trace.	12.00	
Total	152.12	71.39	218.77

Antelope Springs. Twenty miles west of Wagon Wheel Gap, in Antelope Park, are situated Antelope Springs, in a region which is becoming a great resort for sportsmen and abounding in fish and game. The waters of the springs are medicinal and resemble the more widely-known mineral waters of the Gap, in that they are both hot and cold and differ among themselves in their mineral constituents. The scenery is wild and beautiful. For a hunting party, or as a place for a few days' outing in camp, no more pleasing spot can be found.

Trout Fishing in the Rio Grande. There is no stream on the eastern slope of the Rocky Mountains that affords finer trout fishing than the Rio Grande. Trout reaching the wonderful weight of nine pounds have been frequently taken and those weighing from one to three pounds can be caught in great abundance. This is undoubtedly one of the best fishing resorts in America.

ALAMOSA TO ESPAÑOLA AND SANTA FÉ.

HE New Mexico branch of the Denver & Rio Grande Railroad extends southward from Alamosa to Española, passing through an interesting country to the tourist, especially after New Mexico has been entered. Here can be seen what remains of the ancient Spanish civilization, as well as the habitations of the Pueblo Indians and the ruins of the pre-historic Cliff Dwellers. Leaving Alamosa the road turns to the south and crosses the southern portion of the San Luis valley.

EMBUDO, RIO GRANDE VALLEY.

La Jara. Within the last three years many new towns have sprung up in the valley, owing to the development of its agricultural industries, through the construction of great irrigating canals. Old settlements have acquired new vigor and advanced greatly in prosperity. La Jara is one of the towns that has received this new impulse. Its people are enterprising and industrious. Agriculture and pastoral pursuits contribute to the town's success. (Population, 250. Distance from Denver, 265 miles. Elevation, 7,609 feet.)

Manasa. This is the village for a colony of Mormons, which has been established near Antonito. These Mormons do not practice polygamy and are industrious and law abiding citizens.

Antonito. This town is a thriving and prosperous place, the last one of any special importance on the railroad in the southern part of the San Luis valley. Stock raising and agriculture occupy the attention of the surrounding population There is a fine stone depot here, and there are many creditable business blocks. It is the station for Conejos, one mile distant; for Manasa, a large and prosperous Mormon settlement, in which polygamy is not practiced, eight miles distant, and for San Rafael, four miles distant. Its position in the heart of the San Luis valley (for full description of which see Alamosa) insures it a generous and constantly increasing support from agricultural and pastoral industries. Being the junctional point of the Denver & Rio Grande railroad's New Mexico and San Juan branches

OLD CHURCH OF SAN JUAN.

gives it a large railroad business. Tourists will do well to stop at Antonito and visit the old Mexican town of Conejos, which is the most accessible town of the typical Mexican character in Colorado. Here may be found the plazas, churches and ancient adobe houses peculiar to the early civilization of the Spanish. Fine fishing can be found near Antonito. Antonito itself is a modern town with all the life and push of the American, full of business and enterprise. (Population, 250. Distance from Denver, 279 miles. Elevation, 7,888 feet.)

Palmilla is twenty-three miles from Antonito, and here the road enters the Territory of New Mexico and passes through a number of small stations of no especial interest to the tourist. As a matter of statistics, the names of these stations and their distances from Denver are given: Palmilla, 290 miles. Volcano, 297 miles. No Agua, 306 miles. Tres Piedras, 313 miles. Serviletta, 323 miles. Caliente, 335 miles. Barranca, 344 miles. Comanche, 346 miles. Embudo, 351 miles. Alcalde, 359 miles. Chamita, 365 miles. Española, 370 miles. The traveler will notice that the names of the stations have assumed a Spanish form, and should he happen to address any of the swarthy men that chance to be lounging around the sta

A TYPICAL MEXICAN.

tions, he would be very likely to receive a reply in the language of Hispania. The Spanish spoken is not Castilian by any means, but is about as near it as "pidgin English" is to genuine Chinese, being a mixture of English, Spanish and Indian dialects.

OJO CALIENTE.

Famous Hot Springs.

Health and Pleasure Resort.

Elevation, 7,324 feet.

Barranca is a quiet little station in New Mexico, 343 miles from Denver. Its only claim for special mention is the fact that here the traveler takes the stage for Ojo Caliente, the celebrated hot springs, which lie among the hills, eleven miles to the westward. Stages to and from the springs connect with passenger trains, making quick time over an excellent road. The altitude of the springs is about 6,000 feet, and the climate at all seasons of the year mild and pleasant. The springs have been noted for their curative properties from time immemorial, having been frequented by the Indians previous to Spanish occupation and highly esteemed by both races since that date. They have proved remarkably successful in the treatment of rheumatism, skin diseases, derangement of the kidneys and bladder, and especially of all venereal diseases. Cases of paralysis, after resisting the usual appliances of medicine, have been sent to Ojo Caliente, and immediately and permanently relieved. The springs lie in a pleasant valley, one thousand feet lower than Barranca, surrounded by high bluffs capped with basaltic cliffs. On the top of these cliffs are table-lands on which are found the ruins of prehistoric buildings, not unlike the Indian pueblos of the present day, but of which the Indians know nothing and even their traditions furnish no account. Four miles above the village are larger springs of tepid water, the mineral deposits from which have built up great mounds, full of strange caves and glittering with saline incrustations. About three miles from Ojo Caliente is a high mountain called Cerro Colorado, from its peculiar reddish brown color, which, according to the statement of the inhabitants, exhibited marked evidences of volcanic action only fifty-four years ago. It has a well defined crater, and offers an inviting field for the investigations of the geologist.

PUEBLO INDIANS.

Comanche Canon. Six miles below Barranca the train enters Comanche Cañon. Through this cañon the road makes its descent into the Rio Grande Valley. Rugged, difficult and striking, the cañon commands the admiration of the spectator. Through breaks in the walls can be caught glimpses of the valley and river, the noble Rio Grande beneath. Experienced travelers who have made the "grand tour" say that this scene resembles choice bits in Switzerland. Ernest Ingersoll thus describes the valley in his charming book, "The Crest of the Continent": "Emerging from Comanche Cañon, a bend to the southward is made along the western bank of the lower part of the cañon of the Rio Grande. In many portions of this narrow valley, only about twenty miles in length, features of great interest to the eye occur, equalling the walls of Comanche, which was itself ignored until the railway brought it to light. The river here is about sixty yards wide, and pours with a swift current troubled by innumerable fallen rocks. At times it is swollen and yellow with the drift of late rains, but in clear weather its waters are bright and blue, for it has not yet soiled its color with the fine silt which will thicken it between Texas and Mexico. On the opposite bank, near the level of the river, runs the wagon road that General Edward Hatch, formerly commander of the department of New Mexico, cut some years ago to give ready communication between his headquarters at Santa Fé and the posts in the northern part of the Territory and in southern Colorado. This is the track now followed by all teamsters, but the old road from the south to Taos ran over the hills far to the eastward, passing through Picuris."

PUEBLO DE TAOS, NEW MEXICO.

Embudo. At the mouth of Comanche Cañon stands an odd conical hill dividing the current of the river. Noticing its resemblance to a funnel the Mexicans called it Embudo, and the station here takes the same name. Embudo is chiefly important as the point of departure for Taos, whose remarkable pueblo is described further on.

Española. This little village is the southern terminus of the Denver & Rio Grande Railroad, and is of interest to the tourist because of its contiguity to ancient pueblos and the ruins of Cliff dwellings. The Texas, Santa Fé & Northern Railroad connects here with the Denver & Rio Grande Railroad, and by it the journey is continued to Santa Fé. Española's tributary industries are pastoral and agricultural. (Population, 100. Distance from Denver, 370 miles. Elevation, 5,590 feet.)

Places of interest near Espanola.

<pre>
SANTA CRUZ.

PUEBLO OF SAN JUAN.

PUEBLO DE TAOS.
</pre>

Santa Cruz is a most interesting old Mexican town, situated on the Rio Grande del Norte, directly opposite Española. Its chief attraction is the ancient church erected in the sixteenth century, which contains several paintings and images sent over from Spain.

The Pueblo of San Juan is situated on the Rio Grande, about four miles above Española, and one and one-half mile from the railroad There are twenty-six similar Indian towns, nineteen of which are situated in New Mexico, and seven in Arizona. Nine of them are on the line of the Denver & Rio Grande Railway, or its immediate vicinity, viz.: Taos, Picurio, San Juan, Santa Clara, San Yldefonso, Pojuaque, Nombe, Cuyamauque and Tesuque. The

NEW MEXICAN LIFE.

different pueblos closely resemble each other in construction. The dwellings are all built of mud-colored adobes, or sun-dried bricks, and are arranged so as to inclose a plaza or public square. The walls are from two to four feet in thickness, and the roofs are of timbers, covered with dirt a foot or more in depth; many houses are two, and some even four and five stories, or rather terraces, in height, each successive story being set back some twelve or fifteen feet from the side walls of the next story below. The usual manner of entering these dwellings is by ascending a ladder outside the building to the roof, and through a hole descending to the interior by another ladder; though some, as a modern improvement, have doors cut through the side walls. This method was doubtless adopted as a defensive measure during troublesome times, when it was often necessary to convert the pueblo into a fortress from which to repel hostile invasions.

Pueblo of Santa Clara. A few miles below the pueblo of San Juan is the pueblo of Santa Clara, just across the river from Chamita, a station on the Denver & Rio Grande line. Its characteristics are similar to those of the pueblos already described.

The Pueblo de Taos. Twenty miles above Embudo is the Pueblo de Taos. This is considered the most interesting as well as the most perfect specimen of a Pueblo Indian fortress. It consists of two communistic houses, each five stories high, and a Roman Catholic church, now in a ruined condition, which stands near, although apart from, the dwellings. Around the fortress are seven circular mounds, which at first suggest the idea of being the work of Mound Builders. On further examination they prove to be the sweating chambers, or Turkish bath, of this curious people. The largest appears also to serve the purpose of a council chamber and mystic hall, where rites peculiar to the tribe, about which they are very reticent, are performed. The Pueblo Indians delight to adorn themselves in gay colors, and form very interesting and picturesque subjects for the artist, especially when associated with their quaint surroundings. They are skilled in the manufacture of pottery, basket making and bead work. The grand annual festival of these Indians occurs on the 27th of December, and the ceremonies are of a peculiarly interesting character.

All of these ancient pueblos are easy of access via the Denver & Rio Grande Railroad, and abound in objects of interest dating back many hundreds of years before the occupation of the country by the whites, and will fully repay the tourist for the time and expense necessary to visit them.

Española to Santa Fe. At Española the Texas, Santa Fé & Northern Railroad connects with the Denver & Rio Grande and carries the tourist still further southward to the capital of New Mexico, one of the most interesting cities on the North American continent, Santa Fé. En route one can catch a glimpse of the ruins of ancient cliff dwellings perched in the alcoves of the perpendicular bluffs which rise near the track. The journey is only a distance of thirty-eight miles through a country presenting novelty to the eyes of those unfamiliar to subtropical scenes, but not of an especially startling character.

> **SANTA FE,**
> The Oldest town in the United States.
> Commercial City and Health Resort.

The capital of the territory of New Mexico is the oldest city in the United States, there being evidence to show that it was inhabited as early as 1325, or nearly three hundred years before the pilgrim fathers landed on Plymouth Rock. The city of Holy Faith is situated on both sides of the Santa Fé Creek. The streets are narrow, and the buildings are almost all constructed of adobe, and only one story in height. The city is filled with antiquities, the most remarkable of which, perhaps, is the church of San Miguel, built in 1710, and the Palace, erected in 1581. The city is free from malaria and excessive heat and cold, and from wind and sand storms. It is supplied with pure water and pure air from the mountains surrounding; it has delightful scenery beneath bright sunshine with glorious sunsets ; it has trout in its streams, and game in the adjacent hills and mountains ; the people are daily supplied at their doors with the freshest and choicest esculents of home production ; and besides possessing wonderful health-giving properties, it is one of the most comfortable residence cities in the world. This fact is rapidly becoming known and appreciated, as witness its growing popularity both as a summer residence for people from the South, and as a winter residence for people from the North, and as an all-the-year-round residence and sanitarium for people variously in search of health, comfort, pleasure and business.

Santa Fé is the chief money centre of the Territory. It has two old and well established national banking houses, besides hundreds of thousands of dollars for loan in private hands. It is the seat of the general offices in New Mexico of the Atchison, Topeka & Santa Fé Railroad. It has a live board of trade, the most able and distinguished bar in the Southwest. A splendid agricultural, pastoral. and mining country is tributary to the city. (Population, 6,000. Distance from Denver, 408 miles. Elevation. 7,046 feet.)

ALAMOSA TO SILVERTON.

T Antonito the line branches, that to Española and Santa Fé extending due south and that to Silverton turning to the westward. The trip from Antonito to Silverton is one of great interest and abounds in scenic attractions. The road gradually climbs out of the valley of San Luis and up the eastward slope of the Conejos range of mountains. The line from Big Horn to Arboles is constantly among the hills, and the stations are either for the convenience of stockmen or shipping points for lumber, and while of commercial importance to the railroad, of little interest to the tourist. During the summer the Conejos Mountains furnish one of the finest ranges for stock in Colorado, and it goes without saying that these grass-carpeted hills and vales are fully occupied. The forest growth on the western slope is of a larger and more dense character than that of the eastern. Many saw-mills have been here established, and the manufacture of lumber is a large industry. The climb to Chama is full of interest. The line pursues a tortuous course, following the convolutions of the hills and making the ascent up the less difficult grades of the gulches.

Los Pinos Valley. Describing a number of large curves around constantly deepening depressions, we reached the breast of a mountain, whence we obtain our first glimpse into Los Piños Valley, and it comes like a sudden revelation of beauty and grandeur. The approach has been picturesque and gentle in character. Now we find our train clinging to a narrow pathway carved out far up the mountain's side, while great masses of a volcanic conglomerate tower overhead, and the faces of the opposing heights are broken into bristling crags. The river sinks deeper and deeper into the narrowing vale, and the space beneath us to its banks is excitingly precipitous. We crowd upon the platform, the outer step of which sometimes hangs over an abyss that makes us shudder, till some friendly bank places itself between us and the almost unbroken descent. But we learn to enjoy the imminent edge, along which the train creeps so cautiously, and begrudge every instant that the landscape is shut out by intervening objects. To say that the vision here is grand, awe-inspiring, impressive or memorable, falls short of the truth in each case. It is too much to take in at once. We are so high that not only the bottom of the valley, where the silvery ribbon of the Los Piños trails in and out among the trees, and underneath the headlands, but even the wooded tops of the further rounded hills are below us, and we can count the dim, distant peaks in New Mexico.

Phantom Curve. One of the most striking scenes on the line of this ascent is Phantom Curve. Just after the side-track station of Sublette (305 miles from Denver) has been passed, the road makes a great bend around the side of a mountain; on the left rise tall monuments of sandstone, cut by the elements into weird and fantastic figures. Here is indeed a wild spot, with the valley so deep below, the grotesque, red monumental rocks around, the tall, shelving cliffs above. A mile beyond the Curve the railroad crosses the head of the ravine on a high

ALIGNMENT OF TOLTEC GORGE DISTRICT.

bridge of trestle-work. From this point the track runs directly toward the valley, on a line almost at right angles with it, to where it narrows into a mere fissure in the rocks at Toltec Gorge.

TOLTEC GORGE,

A Scenic Wonder.

Depth of Gorge, 1,500 feet.

Distance from Denver, 309 miles.

The approach to this great scenic wonder prepares the traveler for something extraordinary and spectacular. A black speck in the distance against the precipitous surface of a frowning cliff is beheld long before Toltec is reached, and is pointed out as the entrance to the tunnel which is the gateway to the Gorge. As the advance is made around mountain spurs and deep ravines, glimpses are caught of profound depths and towering heights, the black speck widens into a yawning portcullis, and then the train, making a detour of four miles around a side cañon, plunges into the blackness of Toltec tunnel, which is remarkable in that it pierces the summit of the mountain instead of its base. Fifteen hundred feet of perpendicular descent would take one to the bottom of the gorge, while the seared and wrinkled expanse of the opposite wall confronts us, lifting its massive bulwarks high above us,

"Fronting heaven's splendor,
Strong and full and clear."

When the train emerges from the tunnel it is upon the brink of a precipice. A solid bridge of trestle-work, set in the rock after the manner of a balcony, supports the track, and from this coigne of vantage the traveler beholds a most thrilling spectacle. The tremendous gorge, whose sides are splintered rocks and monu-

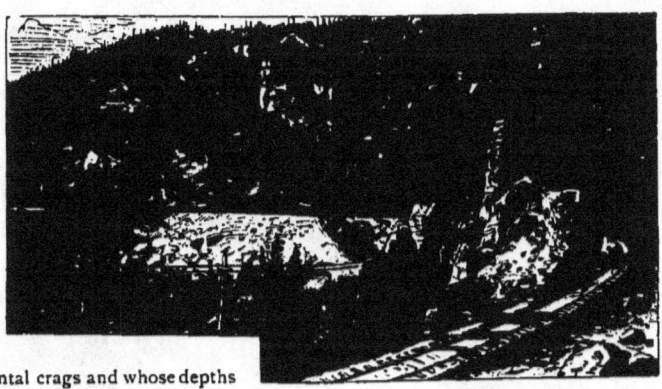

PHANTOM CURVE

mental crags and whose depths are filled with the snow-white waters of a foaming torrent, lies beneath him, the blue sky is above him and all around the majesty and mystery of the mountains.

Garfield Memorial. To the left of the track, just beyond the bridge, stands a monument of granite. Curiosity is naturally excited at beholding this polished shaft, and the questions which arise as to its origin can be briefly answered as follows: On the 20th day of September, 1881, the National Association of General Passenger Agents (then on an excursion over the Denver & Rio Grande Railroad), at the time President Garfield was being buried in Cleveland, held

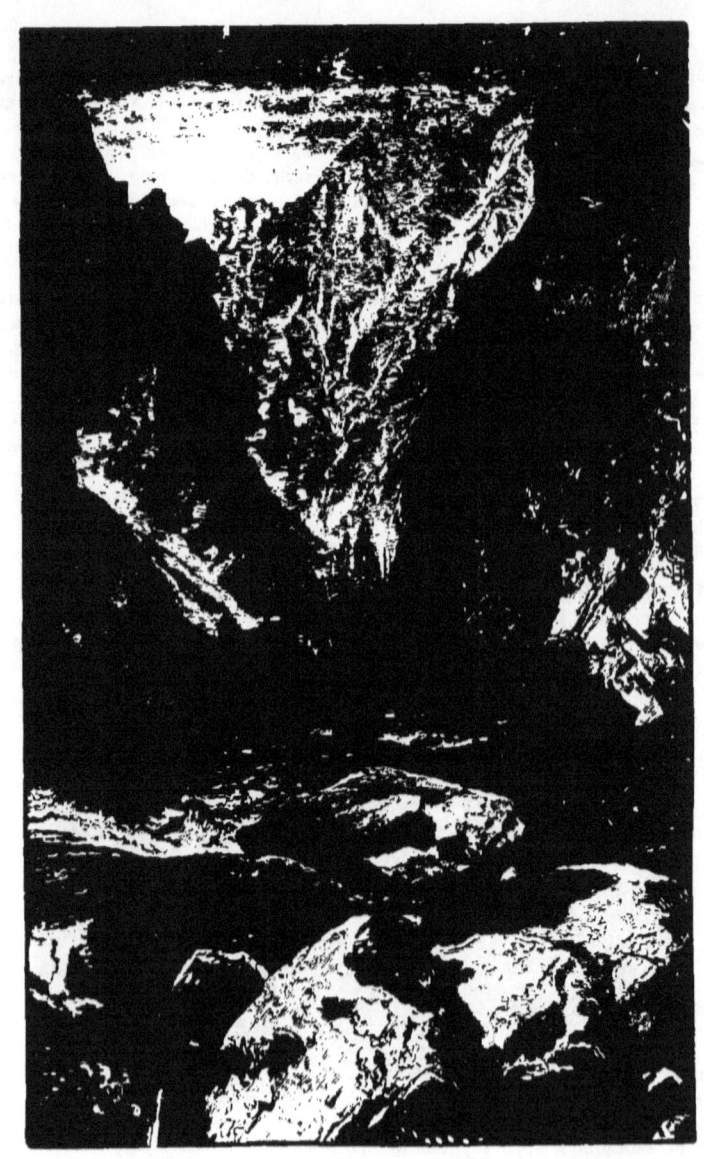

TOLTEC GORGE AND TUNNEL.

memorial services at the mouth of Toltec tunnel and since have erected this beautiful monument in commemoration of the event.

Cumbres. This small station is on the summit of the Conejos Range, which we are now crossing, and, having passed it, we are on the Pacific slope. (Population, nominal. Distance from Denver, 329 miles. Elevation, 10,115 feet.)

Chama. This is an eating station, where, in spite of primitive accommodations, an excellent meal can be obtained. Large quantities of lumber are shipped from here, and the surrounding country is an excellent range for stock. (Population, 250. Distance from Denver, 343 miles. Elevation, 7,663 feet.)

Amargo. This little station is in the midst of attractive scenery, but is especially worthy of mention from the fact that it is the station at which tourists and health seekers take the stage for Pagosa Springs.

GARFIELD MEMORIAL.

PAGOSA SPRINGS.
The "Big Medicine" of the Indians.
HOT SPRINGS,
Health and Pleasure Resort.
Elevation, 7,108 feet.

Pagosa Springs, the far-famed "big medicine" of the Utes, the greatest thermal fountains on the continent, are situated in Conejos County, twenty-eight miles northwest of Amargo, the nearest railway station, on the New Mexico extension of the Denver & Rio Grande Railroad. These springs lie upon the northern bank of the San Juan river, at an altitude of seven thousand feet, and in a situation combining numerous advantages and attractions. To the north are the peaks of the San Juan range east and west are the grassy plains dotted with immense pines and far to the south the undulating prairie stretches into New Mexico. With such an environment, the Pagosa Springs must ere long gain the celebrity to which their medicinal qualities undoubtedly entitle them. The Indians have long been aware of the healing powers of these "great medicine waters," and have, until recently, jealously guarded their possession. It is not surprising that these children of the wilderness, who find relief from distress mainly from the medications of Nature, should deplore the loss of these powerful thermal waters. Within a basin seventy feet long and fifty wide, formed from its own alkaline deposits, which are twenty or thirty feet thick, the water bubbles up at a temperature of 140° Fahrenheit. There are four other springs in the immediate locality, their similarity to the main source, as shown by analysis, suggesting a common origin. Upon a cold morning the steam which rises from these different springs can be seen at a distance of several miles. These purgative, alkaline waters, with the large excess of sulphate of soda, so much increased in medicinal virtue by the degree of temperature, would seem to designate Pagosa as the Bethesda for sufferers from calculus disorders, gravel with uric diathesis, rheumatism

CASTLE OF THE CLIFF DWELLERS, MANCOS CANON, COLORADO

and skin diseases, when alterative and depleting treatment is indicated. New bath houses are being erected, and the tourist will find good accommodations here.

The Pacific Slope. From Chama to Durango, the ride is down grade and through a most interesting country. Hills and valleys of great beauty, meadows covered with thick growing grass, forests of giant trees, are some of the many attractions of this trip. For details of information concerning the small stations, the tourist is referred to the tables given in another part of this book. The line passes through the Indian reservation.

Ignacio. At Ignacio the Indian reservation is reached and the rude tepees of the Southern Utes can be seen pitched along the banks of the Rio de las Florida. Occasionally a glimpse can be caught of a stolid brave, tricked out in all his savage finery, gazing fixedly at the train as it speeds by. Frequently there is quite a little group of these aborigines at the station, and they are always ready to exchange bows and arrows, trophies of the chase, or specimens of their rude handiwork, in return for very hard cash.

DURANGO.
Metropolis
of the San Juan.
Population, 3,500.
Distance from Denver,
450 miles.
Elevation, 6,520 feet.

This thriving city is the county seat of La Plata County, Colorado, and is the commercial centre of southwestern Colorado. It is the market for the agricultural region of Farmington and Bloomfield, New Mexico, and the valleys of the Rio de las Animas, the Rio Florida, etc.

Two miles below Durango is the wonderful "ninety-two feet" thick vein of coal, one of the largest in the State, and here are also great coke ovens. All the surrounding hills are more heavily timbered than in any other part of Colorado. Durango is the station for the whole San Juan mining region, including the following points:

	Population.	Distance from Durango.		Population.	Distance from Durango.
Rico	1500	35	Ft. Lewis, Gov't. Post.	400	12
Ophir	500	62	Farmington, N.M.	500	60
Animas Forks	200	68	Bloomfield, N. M.	200	50
Eureka	100	58	Aztec, Col.	50	30

In a word, Durango is one of the most progressive towns in Colorado, and is surrounded by a country of unexampled richness. Mining, agricultural and pastoral pursuits all contribute to her success; but best of all, her business men are alive, and by their liberality, generosity and push insure a grand future for the city.

THE
CLIFF DWELLINGS.
Relics of
A Pre-Historic Race.
Ruins
Older than History.

One of the most attractive portions of Colorado, to the scientist, antiquarian and, indeed, the general tourist, is that part in which are found the cliff-dwellings of a long extinct race. Some of the most remarkable of these ancient ruins are situated in the Mancos cañon, within a day's ride of Durango. A brief description of one of these will serve as a characterization of all. Perched seven hundred feet above the valley, on a little ledge only just large enough to hold it, stands a two-story house made of finely-cut sandstone, each block about fourteen by six inches, accurately fitted and set in mortar, now harder than the stone itself. The floor is the ledge of the rock, and the roof the overhanging cliff. There are three rooms on the ground floor, each one six by nine feet, with partition walls of faced stone. Traces of a floor which once separated the upper from the lower story still remain. Each of the stories is six feet in height, and all the rooms are nicely plastered and painted, what now looks a dull brick red color, with a white band along the floor. The windows are square apertures with no signs of glazing, commanding a view of the whole valley for many miles. The illustration shows a fortified watch-tower, indicating that these strange cliff-dwelling people were prepared to resist assault. Traditions are few and of history there is nothing concerning this lost race. Their ruined houses only remain, and some broken fragments of the implements made use of in war and peace. Researches are in progress concerning these extremely interesting ruins,

ON THE LOOKOUT.

and new facts are being developed concerning their architecture; but it is quite improbable that any certain light will ever be thrown on their origin or history.

Farmington, Bloomfield and Aztec are growing towns in New Mexico, just over the southern line of La Plata County. They are in the heart of a large agricultural and stock growing district, and near many ruins of the homes of the ancient Cliff dwellers. Between Ophir and Rico are two very large lakes, famous to all dwellers in southern Colorado as fishing grounds. Therein are found thousands of the beautiful and delicious mountain trout, and to the borders of these lakes resort deer, bears and mountain sheep. The altitude of the lakes is 11,000 feet, and they have an area of about ten acres.

Fort Lewis is a military post of seven companies capacity, located on the La Plata River. From there one can see into Utah, and for miles and miles the outlines of the Wasatch Range, 70 to 75 miles distant, loom up.

Trimble Hot Springs are reached nine miles above Durango. The spacious hotel stands within a hundred yards of the road to the left of the track. Here are medicinal hot springs of great curative value, and here, in the season, gather invalids and pleasure seekers to drink the waters and enjoy the delights of

CAÑON OF RIO DE LAS ANIMAS.

this charming resort. The water as it pours out of the rock is at a temperature of 120 degrees, and runs constantly in a stream three inches in diameter. Within two feet of it is another spring flowing as much more in a stream of cold water. Bath houses have been put up, and the hot and cold water can be mixed. The medicinal properties of these springs are beyond question. Four miles further up the Animas valley are the Pinkerton springs of warm water, closely resembling in properties those at Trimble's. Leaving the springs behind, the train speeds up the valley, which gradually narrows as the advance is made, the ascending grade becomes steeper, the hills close in, and soon the view is restricted to the rocky gorge within whose depths the raging waters of the Animas sway and swirl.

ALONG THE ANIMAS RIVER.

Magnificent Scenery. From Durango, the metropolis of the San Juan, to Silverton, the scenery is of surpassing grandeur and beauty. The railroad follows up the course of the Animas River (to which the Spaniards gave the musical but melancholy title of "Rio de las Animas Perdidas," or River of Lost Souls) until the picturesque mining town of Silverton is reached. The valley of the Animas is traversed before the cañon is entered, and the traveler's eyes are delighted with succeeding scenes of sylvan beauty. To the right is the river, beyond which rise the hills; to the left are mountains, increasing in rugged contour as the advance is made ; between the track and the river are cultivated fields and cosy farm-houses, while evidences of peace, prosperity and plenty are to be seen on every hand.

ANIMAS CANON.
A Gem of Beauty.
Depth, 1,500 feet.
Distance from Denver, 470 miles.

This beautiful cañon has characteristics peculiarly its own. The railroad does not follow the bed of the stream, but clings to the cliffs midway of their height, and a glance from the car windows gives one the impression of a view from a balloon. Below, a thousand feet, are the waters of the river, in places white with foam, in quiet coves, green as ocean's depths. Above, five hundred feet, climb the combing cliffs, to which cling pines and hemlocks. The cañon here is a mere fissure in the mountain's heart, so narrow that one can easily toss a stone across and send it bounding down the side of the opposing rock-wall until it falls into the waters of the river rushing through the abyss below. Emerging from this wonderful chasm, the bed of the gorge rises until the roadway is but a few feet above the stream. The close, confining and towering walls of rock are replaced by mountains of supreme height. The Needles, which are among the most peculiar and striking of the Rockies, thrust their sharp and splintered peaks into the regions of eternal frost.

Elk Park is a quiet little nook in the midst of the range, with vistas of meadow and groves of pines, a spot which would furnish the artist many a subject for his canvas. At the end of Elk Park stands Garfield Peak, lifting its summit a mile above the track. Beyond are marshalled the everlasting mountains, and through them for miles extends, in varying beauty and grandeur, the Cañon of the Animas Frequent waterfalls glisten in the sunlight, leaping from crag to crag only to lose themselves at last in the onflowing river. Emerging finally from this environment of crowding cliffs, the train sweeps into Baker's Park and arrives at Silverton in the heart of the San Juan.

SILVERTON.
Picturesque Mining Town.
Population, 2,500.
Distance from Denver, 495 miles.
Elevation, 9,224 feet.

This thriving and picturesque little city is the county seat of San Juan County, Colorado, and derives its support from the surrounding mines, which are scattered in every portion of the county. The output of the camp has swelled from an annual product of $40,000 to $2,000,000 in three years. From 600 to 1,000 tons of ore are shipped weekly from Silverton, and the product is constantly increasing. An industry of no small importance, and which is rapidly assuming large dimensions, is the system of leasing mines, and it may be said that at least one-half of the producing mines are now being worked by lessees. Hundreds of prospects that are in a condition to ship paying mineral are now laying idle, awaiting the arrival of thrifty miners to take and work them under this system. The scenery around Silverton is of the most

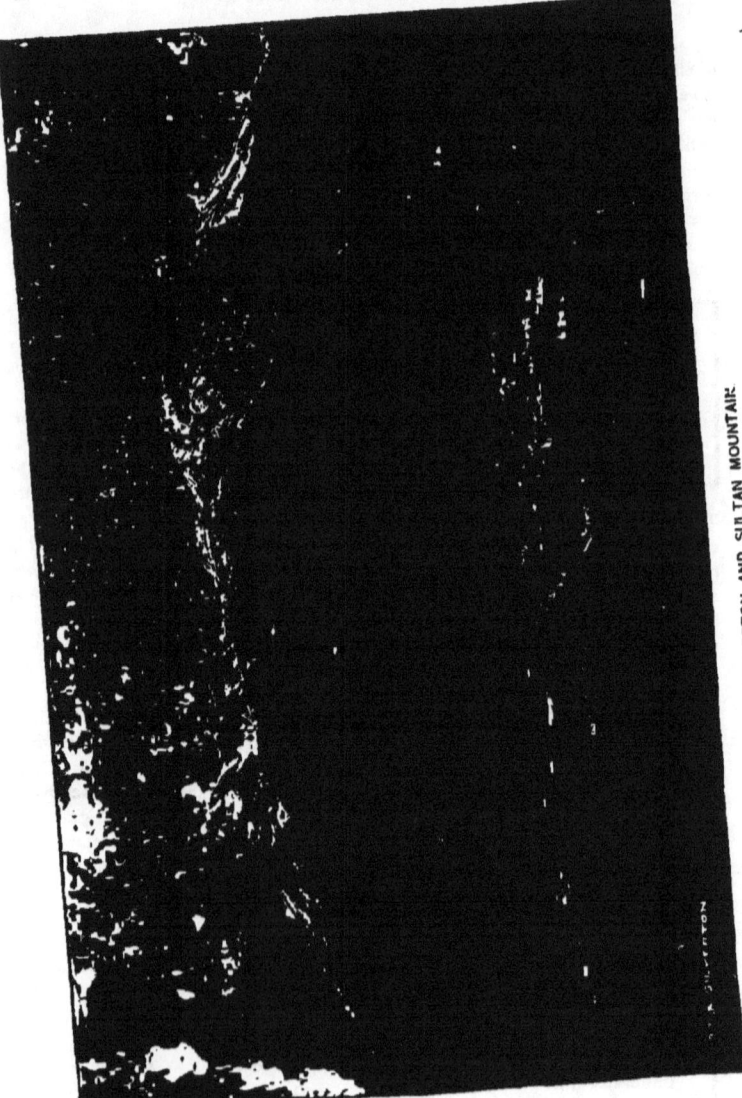

SILVERTON AND SULTAN MOUNTAIN

beautiful and attractive character. Entrance to Baker's Park, in which the town lies, is made through the famous Animas Cañon. Hid in a theatre of hills, the picturesqueness of the surroundings cannot be adequately described. Sultan Mountain, one of the grandest of the San Juan Range, towers above the town ; its summit crowned with snow from which descend innumerable rills, glittering like silver in the sun beams.

ECHO ROCK.

ANIMAS CAÑON AND THE NEEDLE MOUNTAIN.

SILVERTON TO MONTROSE.

HE trip from Silverton to Montrose across the intervening range of mountains, is not at all the difficult undertaking it looks to be. Here, blocking the way, is one of the most rugged and lofty chains of the great Rocky Mountain system, which but recently only the adventurous prospector and his sure footed burro (donkey) dared to cross; but now the journey has been rendered an easy accomplishment by the building of the Silverton Railway over Red Mountain to Ironton, from which point comfortable stages carry the tourist a distance of eight miles to Ouray, where the trip is continued by way of the Denver & Rio Grande Railroad. The construction of the Silverton Railway was a task of great magnitude, and one remarkable feature about it is that it owes its existence to the enterprise and daring of one man. For years Mr. Otto Mears has been the "pathfinder" of the San Juan country, and the toll roads constructed by him have opened the way to the many rich mining camps of that argentiferous region. Recently enlarging his field of usefulness, he began unaided and alone the building of this mountain railroad; himself being both bond owner, stockholder, corporation, president, board of trustees, treasurer, auditor, general manager, chief engineer and paymaster. The result has been one of the most remarkable achievements in engineering of modern times. The road has the same gauge as that of the Denver & Rio Grande, and like it finds no grade so stubborn as to be insurmountable. Taking the cars at the Denver & Rio Grande depot, at Silverton, the ascent of the mountains is at once begun. There is no preliminary skirmishing along level ground for Silverton lies at the bottom of a bowl-shaped valley, and the mountains rise round about on all sides to tremendous heights. With curves, whose sinuosity surpasses that of the serpent's trail, the railroad climbs up the gulches, until at the mining station of Chattanooga the track makes an almost perfect loop, the cars traveling several miles forward and the same distance back — and there lies Chattanooga directly beneath us! All that has been gained is altitude. This is equivalent, however, to a direct progress of a thousand feet, though it has taken a journey of fifteen thousand feet to accomplish it. At the summit of the range the railroad reaches an altitude of 12,000 feet, and the view is something to be remembered a life time. At one point of the descent it has been necessary to construct a switch-back reversing the course of the train, and yet continuing the descent. This switch-back is a novel application of engineering science, and is an exceedingly interesting piece of railroad work. The ascent and descent of Red Mountain by this wonderful railway, give the tourist not only an opportunity to behold the grandest of mountain scenery, but also the privilege of witnessing on all sides the progress of mining operations. The shafts, shaft houses, tunnels, and "prospect" holes of mines in fact or *in futuro*, are to be seen on all sides. The mines of Red Mountain are numerous, and several of them rank among the richest in the world. At Ironton, a typical mining

town, the Silverton Railroad has its terminus, and here stages are taken for the eight miles ride to Ouray.

A Romantic Stage Ride. The stage ride forms one of the most attractive features of this most attractive journey. Lasting only three hours, passing over the summits of ranges and through the depths of cañons, the tourist will find this a welcome variation to his method of travel, and a great relief and recreation. The old fashioned stage, with all its romantic associations, is rapidly becoming a thing of the past. A year or two more and it will have disappeared entirely from Colorado. Here, in the midst of some of the grandest scenery on the continent, the blue sky above, and the fresh, pure exhilarating mountain air sending the blood bounding through one's veins, to clamber into a Concord coach and be whirled along a splendidly constructed road, as solid as the living rock from which it has been carved at an expense in some instances of $40,000 a mile, and as smooth as a city boulevard, is surely a novel and delightful experience. The scenery on

ON THE UNCOMPAHGRE.

this journey between Silverton and Ouray is of the greatest magnificence. This is especially true of this portion of the route traversed by stage. The Silverton and Ouray toll road has long been noted for its attractions in the way of scenery, the triangular mass of Mount Abraham's towers to the left, while the road winds around the curves of the hills with the sinuosity of a mountain brook.

Bear Creek Falls. The scene from the bridge over Bear Creek is one which once beheld can never be forgotten. Directly under the bridge plunges a cataract to the depth of two hundred and fifty-three feet, forming a most noteworthy and impressive scene. The toll road passes through one of the most famous mining regions in the world, and the fame of Red Mountain is well deserved both from the number and richness of its mines. Before Ouray is reached the road passes through Uncompahgre Cañon. Here the roadbed has been blasted from the solid rock wall of the gorge, and a scene similar in nature and rivaling in grandeur that of Animas Cañon is beheld.

OURAY.
The Gem of the Rockies.
Health and Pleasure Resort.
Elevation, 7,721 feet.
Distance from Denver, 389 miles.

This is one of the most beautifully situated towns to be found anywhere. Its scenery is idyllic. The village is cradled in a lovely valley, surrounded by rugged mountains. The situation of the town is thus briefly described in the *Crest of the Continent*: " The valley in which the town is built is pear-shaped, its greatest width being not more than half a mile while its length is about twice that down to the mouth of the cañon. Southward — that is toward the heart of the main range — stand the two great peaks, Hardin and Hayden. Between is the deep gorge down which the Uncompahgre finds its way; but this is hidden from view by a ridge which walls in the town and cuts off all the further view from it in that direction, save where the triangular top of Mount Abrahams peers over. Westward are grouped a series of broken ledges, surmounted by greater and more rugged heights. Down between these and the western foot of Mount Hayden struggles Cañon Creek to join the Uncompahgre; while Oak Creek leaps down a line of cataracts from a notch in the terraced heights through which the quadrangular head of White House Mountain becomes grandly discernable — the easternmost buttress of the wintry Sierra San Miguel. At the lower side of the basin, where the path of the river is beset with close cañon-walls, the cliffs rise vertical from the level of the village, and bear their forest growth many hundreds of feet above. These mighty walls, two thousand feet high in some places, are of metamorphic rock, and their even stratification simulates courses of well ordered masonry. Stained by iron and probably also by manganese, they are a deep red-maroon; this color does not lie uniformly, however but is stronger in some layers than in others, so that the whole face of the cliff is banded horizontally in pale rust color, or dull crimson, or deep and opaque maroon. The western cliff is bare, but on the more frequent ledges of the eastern wall scattered spruces grow, and add to its attractiveness Yet, as though Nature meant to teach that a bit of motion,— a suggestion of glee was needed to relieve the sombreness of utter immobility and grandeur, however shapely, she has led to the sunlight, by a crevice in the upper part of the eastern wall that we cannot see, a brisk torrent draining the snowfields of some distant plateau. This little stream, thus beguiled by the fair channel that led it through the spruce woods above, has no time to think of its fate, but it is flung out over the sheer precipice eighty feet into the valley below. We see the white ghost of its descending, and always to our ears is murmured the voice of the Naiads, who are taking the breathless plunge. Yet by what means the stream reaches that point from above cannot be seen, and the picture is that of a strong jet of water bursting from an orifice through the crimson wall, and falling into rainbow-arched mist and a tangle of grateful foliage that hides its further flowing."

The town has one hotel of great magnificence worthy of a city of ten times its population, besides a good supply of other hostelries of a less splendid character. Ouray is a health resort worthy of patronage by invalids, possessing hot springs of a fine medicinal character and abounding in attractions to divert the mind. Plenty of sport can be had about here. The mountain sheep and wapiti have not yet been killed off; deer and trout are abundant. The rides up the roads and trails to neighboring mines and mining camps, through valley and cañon, and over mountain and mesa, are not soon exhausted, and the lover of botany or geology, or the student of mineralogy and mining, could scarcely find a finer field anywhere than in the neighborhood of Ouray.

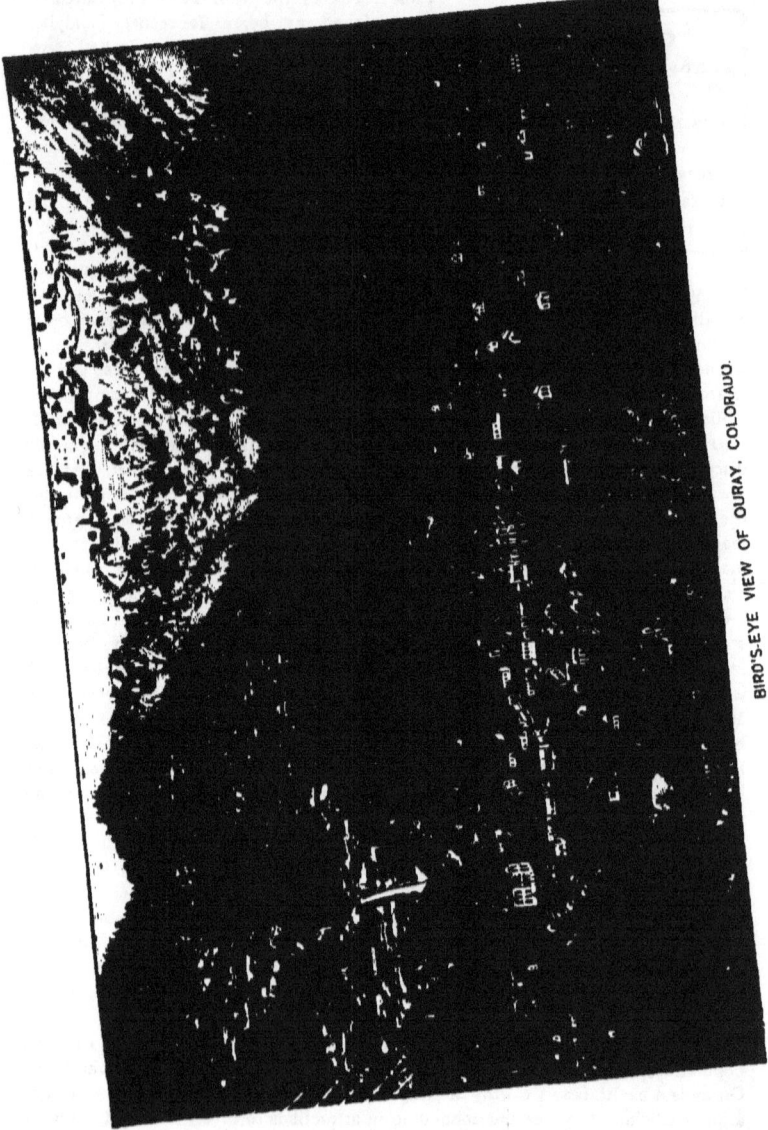

BIRD'S-EYE VIEW OF OURAY, COLORADO.

Ouray to Montrose. Leaving Ouray, a ride of thirty-five miles, via the Denver & Rio Grande Railroad, brings the traveler to Montrose, on the main line of the Denver & Rio Grande Railroad, between Denver and Salt Lake. Two miles from Ouray the country begins to become open and soon one is passing through farms and an excellent agricultural valley. En route one passes the confluence of the Uncompahgre and Dallas, where the wagon road branches to the gold fields of San Miguel, and the mesas and terraces on either side abound with almost every species of game, deer, elk, mountain sheep, bear, and smaller animals. Further on, twenty-two

CHIEFS OF THE UNCOMPAHGRE UTES.

miles from Ouray, you come to the old Los Piños Agency, where Chiefs Douglas, Jack, Colorow, Piah, and other Indians, who participated in the massacre of Thornburg and the Meekers, tested the nerve of General Hatch and his associates in 1879. The store-house, council chamber, etc., are still standing. The military camp is passed twenty-six miles from Ouray, and five miles further on, one reaches the residence of Chippeta, the widow of Ouray, the dead Ute chief, who, during his reign, held the Utes in check, and was always the friend of the white man. At Montrose the tourist can take the main line of the **Denver & Rio Grande Railroad** and resume the trans-continental journey.

BIRD'S-EYE VIEW OF SALIDA.

SALIDA TO ASPEN.

THE trip from Salida to Aspen abounds in interest for the tourist. It leads one through a most varied country, and presents to the inspection of the traveler almost every variety of industry, from the agriculture and stock raising of the Arkansas and Eagle River Valleys, to the gold and silver mining of Leadville and Aspen, and it may be said, in passing, that Leadville and Aspen are the two greatest mining camps in the world and well worthy of a visit. The scenery after Salida is passed grows in interest with each mile of advance. We are steaming up the left bank of the Arkansas River, and are crossing the western border of the

BROWN'S CANON.

Great South Park. The mountains, capped with snow, shut us in throughout the whole circle of the horizon. The Collegiate Range, including the peaks of Yale, Harvard and Princeton to our left, and beyond, the great volcano-made cones of Ouray and Shaveno, which tower above Marshall Pass. Away off to the right are the Kenosha Hills. Agriculture and stock raising are the main industries of South Park, and the ranchmen find these pursuits of an exceedingly lucrative character. A number of small stations are passed beyond Salida as follows: Brown Cañon, Hecla Junction, Nathrop and Midway.

Brown's Cañon. After passing the station of Brown's Cañon, fine views of the Sangre de Cristo peaks present themselves close by, and then the rocks are heaped up again into the grand defile of Brown's Cañon, where one of our illustrations was made.

Calumet Branch. Just before entering Brown's Cañon, a branch road can be seen running off to the northward. That is the short road up to Calumet, where the Colorado Coal and Iron Company have iron mines of great value and in constant operation, for the ore is suitable for the making of Bessemer steel. These mines are open, quarry like excavations, and the ore is therefore more easily handled than is usual. The grade on this branch, four hundred and six feet to the mile, is said to be the heaviest in the world where no cog-wheels are used. Only a few empty cars can be hauled up; and the difficulty is almost as great in descending, for it requires at least four cars, dragging with hard set brakes, to hold an engine under control in going down. Marble and lumber in great quantities are also shipped down this little branch from the neighborhood of Calumet.

Buena Vista. Buena Vista is the county seat of Chaffee County. The town was incorporated in the month of December, 1879, and, for its age, is a wonderfully thriving place. It is beautifully situated on the Arkansas River, thirty-six miles below Leadville and 242 miles from Denver. The town is quite an important station, and is surrounded by good mines of gold and silver, fine pasture-lands for stock and many improved ranches. The city has an abundance of pure water, fine shade trees, churches, schools, stores, etc. (Population, 1,800. Distance from Denver, 242 miles. Elevation, 7,970 feet.)

Cottonwood Springs. The Cottonwood Hot Springs have long been famous in Colorado for their curative properties. They were the resort of the Indians before the whites took possession of the country, and have since been greatly improved and made accessible to invalids and tourists. The springs are situated six miles from Buena Vista, whence a stage line conveys passengers arriving on the Denver & Rio Grande Railroad to the springs. For cases of inflammatory rheumatism, lead poisoning, and diseases of the blood, these waters possess remarkable curative properties. The scenery of the valley in which the springs are situated is of great loveliness, the Collegiate Range of mountains forming an imposing background. Fine trout fishing can be found in ten minutes' walk up and down Cottonwood Creek, and the neighboring hills abound in game. There are good accommodations here for tourists and invalids.

After leaving Buena Vista the following small stations are passed: Americus, Riverside, Pine Creek, Granite and Twin Lakes (station).

TWIN LAKES,
Pleasure Resort.
Elevation, 9,357 feet.

The station of Twin Lakes must not be confounded with the lakes themselves. These most beautiful mountain tarns are best reached by a seven miles stage ride from Granite Station. The drive is in itself a delightful experience, and the lakes prove a most charming culmination. You find yourself in a little valley about seven miles in area. Around you on all sides, looming up grand and precipitous, are snow-capped mountain peaks, each of them towering fully a mile high, from where you stand, completely walling you in from the outer world. These mountains are Mount Elbert, La Plata, and Twin Peaks, each of them higher than the famous Pike's Peak, Lake Mountain, Mount Sheridan, and Park Range. They are all more or less covered, up to the timber line, with fir and spruce trees, the fragrance of which perfumes the atmosphere, and, owing to the rarified air, the tops of the peaks, on which rest the eternal snows, seem so near that you think you could almost throw a stone to their summits, though in fact the length of that very up-hill stone-throw would be considerably more than a mile. For about three-fourths of its area the valley is occupied by the lakes, and to an ordinary observer it is plain that these lakes were formerly one and occupied the whole valley up to the very foot of the mountains. At present, however, they are twins—Siamese twins—for they are connected by a mountain stream, which, as well as the lakes themselves, abound in the most delicious mountain trout that ever nibbled at a hook or smoked on a platter.

Now let us row out into the middle of the upper lake. It seems as if you were in the centre of a mighty amphitheatre, the arena of which is water, the sloping sides fir-clad mountains, and the roof a great bowl inverted, painted a gorgeous blue and lightly resting on the snow-capped mountains. The sizzling dweller of cities may ask what is the thermometer here? I do not know. I never saw one here. These people have no more use for a thermometer than a toad has for a pocket-book. Old Sol rises bright and fierce-looking every morning in an Italian sky, but his rays are so tempered by the breezes from the mountains that by the time they reach the valley they are just pleasantly warm and exhilarating. But there is one thing his rays will do, and city folk would better beware of them if they do not want to peel off their outer cuticle, they will sunburn as effectually as if conveyed through the medium of a burning glass; this is owing to the rarity of the atmosphere. Flannels can and ought to be worn here every day, and a person sitting reading or writing in-doors for an hour or so, in a room where there is no fire, and while the sun is shining brightly outside, will find the cold stealing up his nether limbs.

Returning to Granite and resuming the journey, the following small stations are passed: Hayden, Crystal Lake, Malta and Eiler.

This wonderful Cloud City first became known to fame in 1859 as California Gulch, one of the richest placer camps in Colorado. From 1859 to 1864 $5,000,000 in gold dust were washed from the ground of this gulch. The camp was afterwards nearly abandoned, and it was not until 1876 that the carbonate beds of silver were discovered. Immediately after this discovery a great rush ensued to the carbonate camp, which was named Leadville, and the population rose from a nominal number to 30,000. Leadville is the county seat of Lake County. It is the third city in size in Colorado, and the greatest and most unique carbonate mining camp in the world. The visitor to Leadville is irresistibly reminded of the words of Joaquin Miller : " Colorado, rare Colorado! Yonder she rests; her head of gold pillowed on the Rocky Mountains, her feet in the brown grass ; the boundless plains for a playground ; she is set on a hill before the world, and the air is very clear, so that all may see her well." The city is lighted by gas and electricity ; has telephonic communication with surrounding points ; has the usual conveniences and luxuries of cities of corresponding size, and in all respect ranks as one of the greatest cities of this great State. Leadville is one of the most interesting cities in the world to the tourist. It abounds in scenes of a novel and characteristic nature, and presents views of life entirely foreign to the conventional. Mining methods are here fully illustrated in every form, from lode mining to hydraulic and sluicing work. Leadville has a handsome theatre, the Tabor Opera House, having a seating capacity of 1,000. The scenery around Leadville is magnificent. It is walled in on all sides by towering mountains whose summits are crowned with eternal snow. Occupying so high an altitude, the effect is remarkable, and tourists can find no more striking nor interesting scenes than those

> **LEADVILLE,**
> The Great " Carbonate Camp."
> Population, 20,000.
> Elevation, 10,200 feet.
> Distance from Denver, 277 miles.

presented by Leadville and its weird and wonderful surroundings. Leadville is well supplied with good hotels. Livery accommodations are first-class, and the boulevard affords one of the finest drives in the State. Situated on the front of Mount Mass ve, at the mouth of Colorado Gulch, and distant five miles from Leadville are the popular Soda Springs. The boulevard, a carefully constructed drive, one hundred feet in width, and as smooth as a race track, gives access to the springs, a stage connecting with Leadville twice a day. The springs are strongly impregnated with soda, and are of a highly medicinal character. There is excellent trout fishing within a few minutes walk of the springs, pleasant drives and rides are numerous, and placer as well as lode mining are in progress in near proximity, easily accessible to the inspection of the tourist. As a business point, Leadville is recognized as among the first in the State, with its large population, great smelting works and vast mining industry, it cannot help but command the attention of business men and investors.

Between Leadville and Tennessee Pass are the following unimportant stations: Eagle Junction, Keildar and Crane Park.

Tennessee Pass. Rising along a tortuous path cut at a heavy grade, as usual, into the side hills, we mount slowly into Tennessee Pass, which feeds the head of Eagle River on one side and one source of the Arkansas on the other. It is a comparatively low and easy pass, covered everywhere with dense timber, and a wagon road has long been followed through it. There was nothing to be seen except an occasional pile of ties, or a charcoal oven, save that now and then a gap in the hills showed the gray rough summits of Galena, Homestake, and the other hights that guard the Holy Cross. At each end of the Pass is a little open glade or "park," where settlers have placed their cabins and fenced off a few acres of level ground whereon to cut hay, for nothing else will grow at this great elevation.

MOUNT OF THE HOLY CROSS.

Elevation, 14,176 feet.

We can do no better service to the tourist than to quote Ernest Ingersoll's description of this famous mountain given in "The Crest of the Continent." He says: "One of the side valleys, coming down to the track at right angles from the southwestward—I think it is Homestake Gulch—leads the eye up through a glorious alpine avenue to where the cathedral crest of a noble peak pierces the sky. It is a summit that would attract the eye anywhere,—its feet hidden in verdurous hills, guarded by knightly crags, half-buried in seething clouds, its helmet vertical, frowning, plumed with gleaming snow,—

'Ay, every inch a king.'

"It is the Mount of the Holy Cross, bearing the sacred symbol in such heroic characters as dwarf all human graving, and set on the pinnacle of the world as though in sign of possession forever. The Jesuits went hand in hand with the *Chevalier Dubois*, proclaiming Christian Gospel in the northern forests; the Puritan brought his testament to New England, the Spanish banners of victory on the golden shores of the Pacific were upheld by the fiery zeal of the friars of San Francisco; the frozen Alaskan cliffs resounded to the chanting of the monks of St. Peter and St. Paul. On every side the virgin continent was taken in the name of Christ, and with all the *eclat* of religious conquest. Yet from ages unnumbered, before any of them, centuries oblivious in the mystery of past time, the Cross had been planted here. As a prophecy during unmeasured generations, as a sign of glorious fulfillment during nineteen centuries, from always and to eternity a re-

MOUNT OF THE HOLY CROSS.

minder of our fealty to heaven, this divine seal has been set upon our proudest eminence. What matters it whether we write 'God' in the constitution of the United States, when here in the sight of all men is inscribed this marvelous testimony to his sovereignty! Shining grandly out of the pure ether, and above all tribulence of earthly clouds, it says: Humble thyself, O man! Measure thy fiery works at their true insignificance. Uncover thy head and acknowledge thy weakness. Forget not, that as high above thy gilded spires gleams the splendor of this ever-living Cross, so are My thoughts above thy thoughts, and My ways above thy ways."

Crane's Park is a beautiful park in the mountains at the western foot of Tennessee Pass. Here are to be seen the kilns of charcoal burners, and a wonderful valley and mountain view.

Red Cliff Canon. Just beyond Crane's Park the railroad enters Red Cliff Cañon, a comparatively short but very interesting gorge in the mountains.

Red Cliff. This picturesque little town is the county seat of Eagle County, and the entrepot of a large mining district. The mines of the Battle Mountain and other districts contribute greatly to the business of the place. Leadville with its smelters is only 25 miles distant, and this fact is also an element of success among the many which give promise of future prosperity to the town. The scenery around Red Cliff is of the grandest and most beautiful description. To reach the town the traveler makes the ascent and descent of Tennessee Pass, and obtains the best distant view that can be had of the famous Mount of the Holy Cross. Just beyond Red Cliff are the wonders of Eagle River Cañon (Population, 1,000. Distance from Denver, 299 miles. Elevation, 8,671 feet.)

> **EAGLE RIVER CANON.**
>
> ———
>
> **Height of Walls,**
> **2,000 feet.**

Beyond Red Cliff the Eagle River Cañon opens to the view at first a comparatively wide expanse, later more narrow, walled in on each side by cliffs of vari-colored rocks, whose lofty and apparently insurmountable summits bear the dark banners of the pine. Admiration and awe at this stupendous work of Nature take possession of the mind, when suddenly these emotions are overshadowed by wonder and almost incredulous surprise at the daring of man, for there above us on the right, perched like the nest of heaven-scaling eagles, rest the habitations of men! There are the shaft houses and abiding places of adventurous miners, who, having climbed these cliffs, pick in hand, have here discovered rich veins of the precious metal, which, being blasted from its matrix, is conveyed to the railroad track 2,000 feet below, by a most ingenious system of tramways and endless steel ropes. There is something very impressive in the sight of these frail cliff-perched dwellings; and the shaft-penetrated, tunnel pierced peaks suggest irresistibly the fabled cavernous labyrinths of "Kor." Nowhere can the traveler find a more interesting and instructive illustration of mining methods than is here presented by the shaft-scarred sides of Battle Mountain and the pinnacle perched eyries of Eagle River Cañon.

The Valley of the Eagle. The cañon passed, one enters the Valley of Eagle River. Quieter scenes of pastoral and agricultural achievements follow. Here are comfortable ranch houses surrounded by fertile fields; there are herds of cattle feeding contentedly in natural pastures; while on all sides are seen evidences of peace, prosperity and plenty. The Eagle River, a beautiful stream, whose pellucid waters do not conceal the bright colored gravel of its bed, meanders through the valley, adding to the beauty of the scene, and carrying with it the practical benefits of irrigation, without which the soil would produce nothing but vegetation suitable for grazing purposes. The clear, cold waters swarm with trout, and here the disciples of old Izaak Walton cannot fail to find ample room and verge for plying their gentle craft.

In our journey through the valley we pass the following stations: Rock Creek, Minturn, Allenton, Sherwood, Eagle River, Gypsum, Dotsero, Shoshone and Sulphur Springs.

CANON
OF THE
GRAND RIVER.
A Marvelous Gorge.

One of the World's Wonders.

Gradually the valley narrows, high bluffs hem us in on the left, the river is close to the track on the right, and its fertile banks suddenly change into a tumbled, twisted, black and blasted expanse of scoria, the outpouring of some ancient volcano of tremendous activity. The few trees on the hither side of the stream are also black, an inheritance of fire; the waters under the black banks, and reflecting the blackened trees, take on a swarthy hue — a stygian picture! Just beyond, a distant glimpse of fertile country, and the clear waters of the Eagle are lost in the muddy current of the Grande, and a cañon greater in extent and more varied in character than that of the Arkansas opens before us. As the train speeds

EXPLORING THE WALLS.

downward, the mountains on the horizon behind us seem to rise up towards the zenith as though the miracle of creation was being repeated before our eyes. Soon, however, the distant mountains are shut out and only the sky above, the river and track beneath and the cliffs around are visible; and here begins a panorama, kaleidoscopic in its ever changing forms and colors, the wonder of the one who sees, the despair of the one who wished to tell others what he saw.

In places the effect is that of giant Egyptian art and architecture. Vast bastions of granite, strata on strata, rise to a stupendous height, braced against rock masses behind them, infinitely vaster. Suggestions of the Sphinx and of the pyramids can be caught in the severe and gigantic rock-piled structures on every hand. These are not made up of boulders, nor are they solid monoliths, like those in the Royal Gorge. On the contrary, they are columns, bastions, buttresses, walls, pyramids, towers, turrets, even statues, of stratified stone, with sharp cleavage, not in the least weather-worn, presenting the appearance of Brobdignagian masonry— hence I use the phrase "rock-piled structures" advisedly and as best descriptive of what there exists.

But the kaleidoscope is shaken and the rock pieces are re-arranged. The effect is startling. We have left Egypt, with her shades of gray and her frowning, massive and gigantic forms. We are in a region of glowing colors, where the vermilion, the maroon, the green and the yellow abound and mingle and contrast. What strange country was the prototype of this? Ah! yonder is something characteristic—a terraced pyramid banded with brilliant and varied colors—the teocoli of the Aztecs.

Whirling around a headland of glowing red rock, which it seems ought to be called "Flamingo Point," we are in a region of ruddy color and of graceful forms. Minarets, from whose summits the muezzin's call might readily be imagined falling upon the ears of the dwellers in this "Orient in the West," spires more graceful than that of Bruges, more lofty than that of Trinity, towers more marvelous than Pisa's leaning wonder, columns more curious than that of Vendome, splintered and airy pinnacles, infinite in variety, innumerable! inimitable! indescribable!

In a moment darkness and the increased rumble of wheels; then light and another marvelous view. We have passed tunnel No. 1, the portcullis; darkness again for a moment, then the blue sky above us. We have entered through the postern gate; darkness for the third time—absolute, unmitigated blackness of darkness; this must be "the deepest dungeon 'neath the castle moat." But soon again we see the blessed light, and there before us lies the goal of our journey— Glenwood Springs.

GLENWOOD SPRINGS,
Health and Pleasure Resort.
Wonderful Hot Springs.
Distance from Denver, 367 miles.
Elevation, 5,768 feet.
Population, 3,000.

Glenwood Springs is the pleasure and health resort of western Colorado, as well as a flourishing and growing town. It is the county seat of Garfield County. The picturesque scenery of the Grand River, from its source midst the peaks and crags of the Rockies, to its debouch into the magnificent waters of the broad Colorado, has been the theme of able writers in prose and poetry, but at no spot in its rapid march to the sea, do the waters of the Grand glisten and ripple upon the shores of a lovelier valley than at its confluence with the Roaring Fork, where are situated the springs and city of Glenwood. Here the sentinel ranges, which have guarded the stormy passage of the turbulent stream through mountain pass and precipitous cañon, seem to have deployed their ranks, that they might surround and embrace a valley so lovely in its landscape and set in a frame of such scenic grandeur. The springs themselves are phenomenal, innumerable fountains bubbling up over an area covering both sides of the river, and varying in volume from twenty to one thousand cubic inches per second. The principal springs on the north side of the Grand River discharge an immense body of water, heated in nature's furnace to 140 degrees Fahrenheit, which flows in a broad stream to its outlet through an aqueduct recently constructed, forming a beautiful island, upon which is erected a commodious and well appointed bathing house, provided with every convenience for sitz, plunge and vapor bathing. The waters have been found of great benefit to invalids, and as a result the springs are largely patronized. Aside from the beautiful valley selected from its site, and the attractions presented by its wonderful springs, Glenwood City possesses many advantages and material resources which are destined to make it one of the most important points on the Denver & Rio Grande Railroad. The town has electric lights, water works, and most of the modern improvements. Good hotels provide for the comfort of the tourist. An illustration of the enterprise of these people is the fact that all the material for the first hotel erected here was brought in over the mountain ranges on the backs of burros and by mule trains. Situated in the midst of a vast agricultural and stock growing region, the tide of immigration is rapidly filling the valleys and uplands with actual settlers, whose traffic has built up for the town during the past year a large trade, which is only a suggestion of the vast mercantile traffic which will be done at this central point in the near future, when this inviting section of the great West is populated by the immense number of inhabitants it is capable of sustaining.

Accommodations for Bathing. The bath house recently erected at the wonderful hot springs here, is of the most elegant design. It is built of red sandstone, and the walls of all rooms are of red or cream colored pressed brick, wainscoted with Texas pine and colored enamels. There are forty-four large bath rooms, in two departments, for the respective sexes. Each bath room has two compartments. One is lined with enamel and set with a porcelain tub, having bronze appliances for readily supplying hot, warm or cold, mineral water; and hot, warm or cold, fresh water, also showers of warm or cold water. Any desired temperature, from 45° up to 120° Fahrenheit can be supplied. The other compartment is furnished as a dressing room, and provided with a settee for reclining after the bath. These compartments have high ceilings and are well lighted from elevated windows by day, and by incandescent electric lamps at night. Light refreshments are served in each room by attendants summoned by electric bells. Massage treatment is administered in a room for that purpose. Besides the bath rooms, the building contains handsome sitting and smoking rooms with open fires, physician's room. billiard room, coffee kitchen, linen rooms, hair dressing rooms, laundry, etc. All rooms are kept supplied with fresh air at an equable temperature throughout the year. Every accessory for the luxurious and health-giving bath is provided in the building. The baths are supplied from the main spring, which yields a constant flow of 2,500,000 gallons per day of highly mineralized hot water, at a temperature of 124.2° Fahrenheit. This water is a remarkable remedial agent, aiding or effecting cures of scrofula, rheumatism, gout, lead poisoning, diabetes, Bright's disease, and all skin and blood diseases. The new bath house stands on the margin of the Mammoth Swimming Pool.

The Bathing Pool. This is remarkable for its size and the completeness of its conveniences. It is nearly six hundred feet in length, by one hundred and ten feet in width at the widest part. Its depth gradually increases from three and one-half feet at one end to six feet at the other. The walls are of red sandstone, and the bottom is paved with hard pressed brick. Its surface area is 43,000 square feet, or one acre; and the capacity, 1,500,000 gallons. It is constantly supplied with mineral water from the main and Yampa Springs, and kept at a temperature of about 95° Fahrenheit. There are one hundred and thirteen dressing rooms, in separate departments for the sexes. These are warmed in winter, and a hooded way leads into the water, so that bathers use the pool with safety and comfort in mid-winter. At night the pool is brilliantly lighted by arc electric lights. Bathing suits are supplied at a moderate charge. Thousands who have tried bathing in the pool pronounce it the most delightful of baths. The exercise which it admits of while bathing is deemed especially beneficial to many kinds of invalids.

The Vapor Caves. A remarkable feature of these springs are the vapor caves—natural openings in the rocks to which the steam from the hot springs obtains access. In one of these natural caves the company has erected a unique vapor bath house with ample dressing rooms, a number of private vapor rooms, shower bath room, etc., all lighted by electric lights, affording vapor baths in either cave or private rooms at a temperature of 105° to 110° Fahrenheit. These baths are not only a luxury to those who are well, but are especially recommended by physicians for a number of serious ailments.

Extension of the Denver & Rio Grande Railroad. From Glenwood Springs an extension of the Denver & Rio Grande Railroad is in process of construction, which will be of great importance to the traveling public. It has been built as far as New Castle and will be extended to Grand Junction, passing down the Grand River and through a continuation of the marvelous Cañon of the Grand. When completed it will become part of the great trans-continental line, as Ogden can then be reached either via Salida and Marshall Pass, or via Leadville and Tennessee Pass.

Carbondale. Situated at the confluence of Rock Creek and Roaring Fork, twelve miles south of Glenwood Springs. This is the proposed point for coking ovens and blast furnace to be erected by the Colorado Coal & Iron Company. (Population, 500. Distance from Denver, 379 miles. Elevation 6181 feet.)

The Elk Mountain Railway, starting at Carbondale, where it connects with the Aspen branch of the Denver & Rio Grande Railroad, runs up the valley of Rock Creek, in a southerly direction, for about twenty-two miles, and then in nearly an easterly line to Robinson's Lake, thirty miles from Carbondale. The line will be constructed from Robinson's Lake to the mines in the vicinity of Crystal, in the near future, making a distance of thirty-five miles from Carbondale.

Avalanche Creek. Twelve miles from Carbondale. This will be the shipping point for silver and iron ores located six to eight miles up Avalanche Creek.

Penny's Hot Springs. Fourteen miles south of Carbondale, on Rock Creek. These springs are said to be equal to those of Glenwood in healing and restorative power.

Coal Basin. Nineteen miles from Carbondale. At this station all the coal from Coal Basin will be received. This is the largest and finest body of coking coal in Colorado, and is largely controlled by the Colorado Fuel Company. Extensive coking ovens will be erected at this point.

Prospect. Twenty-one miles from Carbondale. At or near this point will be located the coal breaker and extensive plant of the Pacific Coal & Coke Company, who own the extensive anthracite coal fields of Chair Mountain. This company is

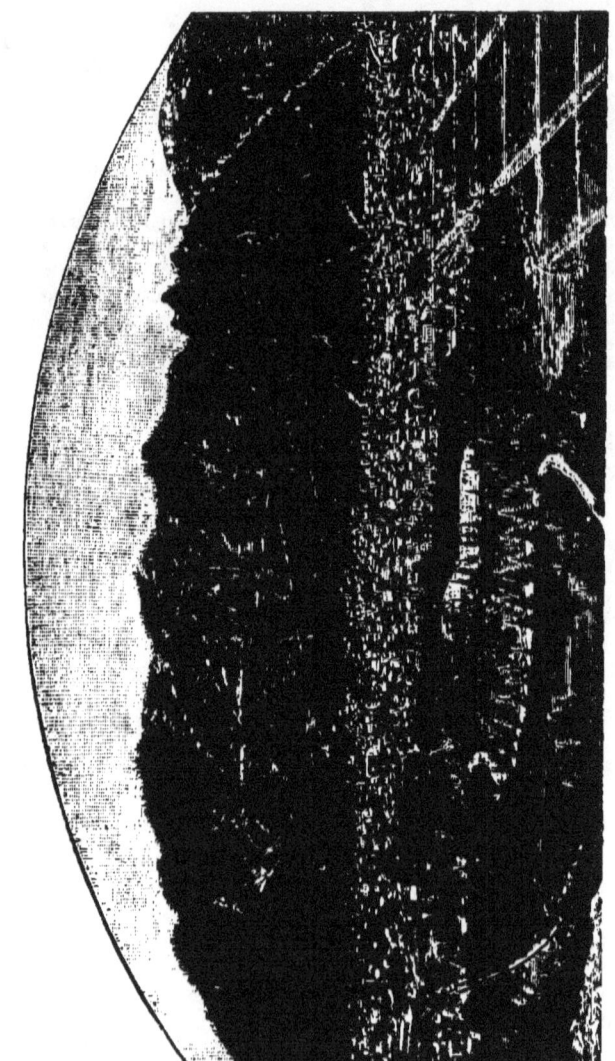

BIRD'S-EYE VIEW OF ASPEN.

preparing to ship five hundred tons of anthracite coal per day. This coal is said to be equal to the best red ash coal of Pennsylvania.

Robinson's Lake. Thirty miles from Carbondale. At this point are located the finest marble and slate quarries west of the Missouri River. Yule Creek joins Rock Creek here, and all the valuable silver ores of that district will be loaded here.

Crystal. This is an old mining camp and is thirty-five miles southeast from Carbondale, on the head waters of Rock Creek. In the vicinity of this camp are located nearly one hundred and fifty patented silver mines. It is estimated that the output from this district will be one hundred tons per day.

Scenic Attraction. The line passes the base of Sopris Mountain and Chair Mountain, and terminates in the great elbow of the Elk Mountains at Crystal. No finer scenery can be found in the West. In a ride of two hours the tourist can be transported from the beautiful valley of Roaring Fork nearly to the summit of the Elk Mountain Range, and can view nearly all the prominent peaks from Mount Massive west.

Returning to Carbondale, the stations on the main line to Aspen are as follows: Emma, Snow Mass and Woody Creek.

ASPEN.
Great Mining Town,
Health and Pleasure Resort.

Aspen, the county seat of Pitkin County, is located in one of the most noted mining regions of Colorado seventy-five miles northwest from Buena Vista, and is the terminus of the Aspen extension of the Denver & Rio Grande Railroad. The valleys of the Roaring Fork River and its confluents, Castle, Hunter's and Maroon Creeks, are especially fitted for agriculture, and the hills and mesas adjacent form a fine range for stock, which in addition to the mining interests will surely make Aspen one of the most prosperous cities in the State. Stores and shops of all kinds, carrying large lines of goods, are abundant, and the business done here would do credit to a town boasting five times its present population. The good faith of the people is manifested by the character of the buildings they have erected. It is a town of beautiful homes, and has most excellent society. All the principal religious denominations have suitable houses of worship, and the public schools are of an excellent order. The hotels are good, there is a fine opera house, and the town is supplied with pure water from Castle Creek. An electric light plant illuminates the principal places of business as well as the streets. The climate is delicious and especially beneficial in all pulmonary complaints. Aspen is a garden town, and displays many beautiful lawns, sprinkled and beautified by flowers.

The main industry of Pitkin County, of which Aspen is the county seat, is mining. The town is situated upon the great zone or belt which passes through the country in a northeasterly and southwesterly course, and has tributary territory for from twenty to thirty miles each way. The ores are of good grade and are found in remarkably large deposits. The Great Central lead with its spurs and lateral feeders, resembles a river with many branches. Silver and lead are the principal mineral products, although gold has been found and profitably worked at Independence, in the eastern part of the county, and the iron ores at Cooper's Camp, in the southwestern part, are found in immense deposits, and are of the very finest quality. Building stone is found, and the rock is unsurpassed in texture or color, and the surrounding hills will be great producers for outside markets. Some coal is found in Pitkin County, but not in extensive measures as in Garfield, the great coal county of the United States, which adjoins upon the north. There is no territory of similar area with richer or more varied products than Pitkin County. The scenery around this thriving city is wonderfully varied and beautiful. Situated in the heart of the mountains, and surrounded by the most wonderful works of nature, Aspen will always be an attractive place to the tourist and the lover of the grand and marvelous. Hunting and fishing are found here in their perfection. Nature seems to have made Aspen her favorite child, and has poured out at her feet all the rich gifts of her cornucopia. (Population, 6,500. Distance from Denver, 408 miles. Elevation, 7 868 feet.)

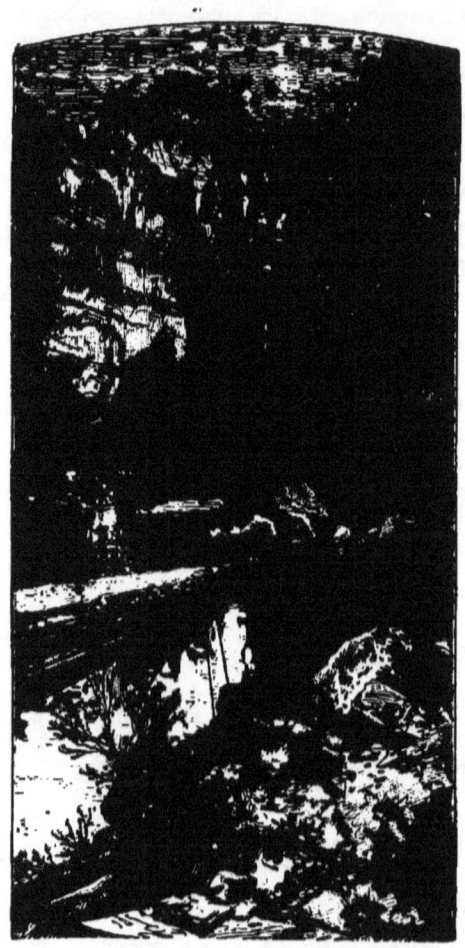

MARBLE CAÑON.

LEADVILLE TO DILLON.

A ROCKY MOUNTAIN BEAUTY SPOT.

From Leadville a branch line of the widely radiating Denver and Rio Grande system extends over Fremont Pass to Dillon. The general direction taken by the line is to the northeast, with a deflection from Frisco to Dillon to the northwest. The Great Middle Park of Colorado lies to the north of Dillon, just over the range of the Williams River Mountains. The country between Leadville and Dillon is extremely mountainous, and mines of great value have been discovered in this region. The railroad crosses the Park Range at Fremont Pass, and in the valley at the foot of the pass the Arkansas River has its sources. The Blue River heads on the Pacific slope near the pass, and the south branch is crossed by the railroad near the small station of Wheeler, the north branch is encountered at Frisco in the vicinity of which the two join and form the main stream, which empties into the Grand in the southwest corner of Middle Park.

The ride from Leadville to Fremont Pass is one of great interest to lovers of the grand and beautiful in nature. The mountain ranges which surround the "Carbonate Camp" are in plain view, and every turn in the road reveals new attractions. This extension of the line is known as the Blue River branch. It is thirty-six miles in length, with its terminus at Dillon. The intervening stations are Birds Eye, Alicante, Fremont Pass, Robinson, Kokomo, Wheeler, Frisco and Dillon.

Source of the Arkansas. The line from Leadville follows up the Arkansas River, and here we have an object lesson in the growth of rivers. We see from what small beginnings great things in the way of water courses grow. We see how a little brook which one could dam with a couple of shovels of mud may push its way along, "undermining what it cannot overthrow; sliding around the obstacle that deemed itself impassable, losing itself in willowy bogs, tumbling headlong over the error of a precipice or getting heedlessly entrapped in a confined cañon; escaping down a gorge with indescribable turmoil, and always growing bigger, bigger, broader and stronger, deeper and more dignified; till it can leave the mountains and strike boldly across a thousand miles of untracked plain to 'fling its proud heart into the sea.'"

Almost in the very springs of the river, where an amphitheatre of gray quartzite peaks stand like stiffened silver-gray curtains between the Atlantic and the Pacific, we curl round a perfect shepherd's crook of a curve, and then climb its straight staff to the summit of Fremont Pass.

> **FREMONT PASS.**
> One of the Highest
> Railroad Passes
> in the World.
> Elevation, 11,329 feet.

Through a charming valley the approach to Fremont Pass is made. A famous pass, with the historic name of him who has been called "The Pathfinder," although a later day has witnessed greater achievements than his among the Rocky Mountains. A journey here deserves the title of a pilgrimage, for from the summit of this pass the traveler can discern the Mount of the Holy Cross. The scene is one replete with vivid interest. Fainter and fainter grow the lines of objects in the valley, until at last the clouds envelope the train, and at the next moment the observer looks down upon a rolling mass of vapor through which the light strikes in many colored beams. The sublimity of the scene forbids all thoughts other than those of reverence and rapture.

> "The snow-crowned monarchs of an upper world,
> Rugged and steep and bare, the mountains rise;
> Their very feet are planted in the skies;
> Adown their sides are avalanches hurled.

> "Time was when few and daring were the men
> Who might behold this pass, that Fremont gained
> Through toil and danger, and its heights attained,
> Perils beset the long leagues down again.

> "Now all may come who seek, afar from crowds,
> The grand in nature, for we now engage
> The potent genii of this iron age,
> Fire, steam and steel, and rise above the clouds!"

The railroad crosses the pass at an elevation of about two miles above the level of the sea, and ranks among the highest railroad passes in the world.

Mount of the Holy Cross. From the crest of Fremont Pass the traveler looks eagerly about and soon catches sight of the sacred symbol which gives name to the famous mount. The snow-white emblem of Christian faith gleams with bright splendor against the azure sky. The wayfarer at last realizes that he has reached the height "around whose summit splendid visions rise." This is one of the best points of view from which to behold this wonderful mountain, a more extended description of which will be found in the chapter

FREMONT PASS.

entitled, From Leadville to Aspen.

Downward to Dillon. On the Pacific Slope are the mines which made this region famous. The Robinson Consolidated, the White Quail, the Wheel of Fortune, etc

Moving on down the pleasant valley, whose level bottom is carbonate tinted, not with ore dust, but with an almost continuous thicket of stunted red willows, we pass the Chalk Mountain mines, the Carbonate Hill district, Clinton Gulch, where gold ore is alleged to be worth more attention than it is receiving, and so come to Elk Mountain and Kokomo. The ore found here is a hard carbonate, running about twenty-five ounces in silver and twenty-five per cent. in lead, besides a third of an ounce in gold, which is carefully separated at the smelter. Much of it is so admirably constituted that it "smelts itself,"— that is, it requires little or no addition of lead, iron and other accessories to its proper fluxion. Continuing the journey we behold alluring pictures of mountains and cañons, of belts of timber and pleasant uplands, of green meadows and sparkling streams beloved of gamey trout and the haunts of deer and elk. This country is still a paradise for the sportsman, and the rod and gun find ample range for their employment here.

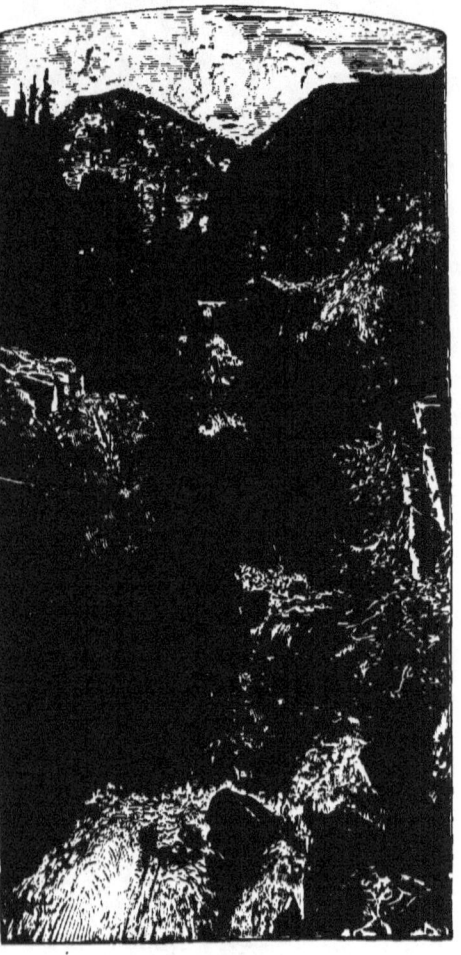

CASCADES OF THE BLUE

Dillon is the terminus of the Blue River branch, and is situated in a mining country. Distance from Breckenridge, 10 miles; Frisco, 3 miles; Montezuma, 12 miles; Decatur, 15 miles; Rock Creek, 10 miles. The station is the nearest point for the lower Blue River Valley, into which good roads extend. Saddle horses and wagons can be hired to go down this river into the hunting and fishing grounds of Middle Park. (Population, 200. Distance from Denver, 313 miles. Elevation, 8,861 feet.)

OGDEN TO SAN FRANCISCO.

OGDEN,
Railroad and Manufacturing Town.
Population, 15,000.
Elevation, 4,286 feet.
Distance from Denver, 771 miles.
Distance from San Francisco, 883 miles.

At Ogden the tourist steps from the train of the Denver & Rio Grande Railroad into that of the Southern Pacific, the transfer being made with very little trouble, and in a few minutes time, at the Union Depot. A glance around will show one that Ogden is beautifully situated on the west slope of the Wasatch Mountains. It is well laid out and substantially built; the streets are wide, regular, lined with shade and ornamental trees, and lighted with electricity By a good system of water works the mountain streams and springs are made to supply an abundance of pure water. Many of the private residences and grounds are very handsome, and the business blocks solid and elegantly constructed. Of the climate too much cannot be said. Utah claims the finest climate in the United States. Colorado makes the same assertion; so does California. There is no doubt that each of these great commonwealths has good grounds for its claims. Colorado and Utah have similar characteristics, while California is quite different; circumstances are said to alter cases, and this saying holds true in climate as well as in other matters. While the climate of Colorado or Utah might be a specific for one class of diseases, that of California might be much more beneficial for another class. The advice of an intelligent and unprejudiced physician should be taken before an invalid decides on his choice of location. In Utah the winters are short and mild, and the spring and fall months give almost perfect weather; the summers are warm but not oppressively hot, and the nights are always cool and never moist. Pulmonary troubles will surely find relief, and generally a cure. Ten miles north of Ogden are Hot Springs, whose sulphur water possesses peculiar medicinal properties, and are pronounced superior to the Arkansas Springs. Hundreds of invalids visit these springs annually, and they are steadily growing in popularity. The educational and religious advantages of Ogden are on a par with those of eastern cities of the same size. Here is the centre of one of the richest agricultural and mining districts of Utah. Ogden has better railroad facilities than any other town in the territory. It is affectionately called by its inhabitants the "Junction City of the West." It is the terminus of five leading trunk lines, namely: The Denver & Rio Grande, the Union Pacific, the Southern Pacific, the Utah & Northern, and the Utah Central Railroads. The outlook for manufacturing is excellent, the Weber River furnishing almost unlimited water power. Iron ore is found in great quantities in the near vicinity, while the wool clip of the territory, and those of Idaho, Wyoming, Montana and Nevada is enormous, and could be advantageously manufactured into cloth at this point.

Geological Features. Looking from the car window after passing

Ogden, the traveler can see many things in this region indicating a thrilling geological history. That striation, extending along the side of the foot hills to the right, marks the water line of a vast, pre-historic inland sea, that shrunk ages ago to the comparatively small proportions of Salt Lake. In all probability the whole area between the Wasatch Mountains and the Sierra Nevadas was once an immense body of water, in which the mountain ranges rose as islands. The lakes of the present day are all that remain of this vast pre-historic sea. The deposits which cover the low lands are chiefly calcareous, and are often filled with fresh water and land shells, indicating a comparatively modern origin. The formation of the islands and the shore ranges of Salt Lake is metamorphic, the strata are distinctly marked and highly inclined, but attaining no great elevation, being generally overlaid with sandstone and limestone of the carboniferous age, but partly altered, the former constituting the loftier eminence, in places it is rich in fossils, while in others it loses the granular character, and becomes sub-crystalline or threaded by veins of calcareous spar, the sandstones, from metamorphic action, taking the character of quartz. As the train advances, evidences of volcanic action become numerous.

Brigham. A half hour's ride from Ogden brings the traveler to Brigham, a busy little town surrounded by an agricultural settlement, but possessing nothing of special interest to the tourist. (Population, 1,800. Distance from Ogden, 17 miles. Elevation, 4,229 feet.)

Corinne. Between Brigham and Corinne the Bear River is crossed by a bridge twelve hundred feet in length. The town of Corinne has a good agricultural country around it, and wherever irrigation has been secured large crops have responded to industrious cultivation. The raising of stock, is also a tributary industry, and cattle do well on the surrounding excellent ranges, which are found in the greatest perfection north of the town. (Population, 500. Distance from Ogden, 24 miles. Elevation. 4,231 feet.)

PROMONTORY.

A Point of

Historical Interest.

A small station surrounded by country covered with sage brush, and only worthy of mention for its history. At this point, on Monday, May 10, 1869, the Union Pacific Railroad, building west and the Central Pacific Railroad, building east, met. The junction was made, and the news flashed all over the world that the first great trans-continental railroad of America had become an accomplished fact. The importance of that event cannot be overestimated, and to enumerate the results emanating from that meeting would be the task of a historian. An epitome of what that meeting meant can be best expressed by quoting that clever and quaintly humorous poem, written by Bret Harte, commemorative of the occasion, under the title of

WHAT THE ENGINES SAID

What was it the Engines said,
Pilots touching—head to head,
Facing on the single track, ,
Half a world behind each back?
This is what the Engines said
Unreported and unread :

With a prefatory screech,
In a florid Western speech,
Said the Engine from the West

"I am from Sierra's crest;
And if altitude's a test,
Why, I reckon, it's confessed,
That I've done my level best"

Said the Engine from the East:
"They who work best talk the least.
S'pose you whistle down your brakes;
What you've done is no great shakes,—
Pretty fair,—but let our meeting
Be a different kind of greeting.
Let these folks with champagne stuffing,
Not their Engines, do the *puffing.*"

"Listen! Where Atlantic beats
Shores of snow and summer heats;
Where the Indian autumn skies
Paint the woods with wampum dyes,
I have chased the flying sun,
Seeing all he looked upon,
Blessing all that he has blest,
Nursing in my iron breast
All his vivifying heat,
All his clouds about my crest;
And before my flying feet
Every shadow must retreat."

Said the Western Engine, "Phew!"
And a long, low whistle blew.
"Come now, really, that's the oddest
Talk for one so very modest,—
You brag of your East! *you* do?
Why, I bring the East to *you!*
All the Orient, all Cathay,
Find through me the shortest way,
And the sun you follow here
Rises in my hemisphere.
Really,—if one must be rude—
Length, my friend, ain't longitude."

Said the Union, "Don't reflect, or
I'll run over some Director."
Said the Central, "I'm Pacific,
But, when riled, I'm quite terrific,
Yet, to-day we shall not quarrel,
Just to show these folks this moral,
How two Engines—in their vision—
Once have met without collision."

That is what the Engines said,
Unreported and unread;
Spoken slightly through the nose,
With a whistle at the close.

Monument. Before Monument is reached the side track stations of Rozel and Lake are passed. At Rozel, the great Salt Lake is close to the track on the left, and at Monument, a point of the same name extends into the lake. Here we take our last view of the interesting and mysterious sea which has been our almost constant companion since leaving Salt Lake City. Before us stretches a vast unfertile country, and here, if anywhere, can be found that makeshift of the easy going and old fashioned geography — the "Great American Desert."

Kelton. This little place is situated on the eastern edge of the desert,

OVER THE RANGE.

TO THE GOLDEN GATE.

and here tne water-trains of the railroad company ootain their supply of the aqueous fluid and deliver to the stations to the westward on this division. Looking to the north the traveler will see the Red Dome mountains, while to the southeast rises Pilot Knob, a prominent feature in the landscape. (Population small. Distance from Ogden, 92 miles. Elevation, 4,222 feet.)

Towns in the Desert. From Kelton to Toano the road traverses the northern edge of the desert, amidst a scene of general desolation. In a general way this unfertile region may be described as sixty square miles of alkaline sands, evidently a portion of the great ocean bed already referred to. Like the arid country, between Fruita and Green River, in Utah, through which we came, on the Denver & Rio Grande Railroad, which only needs irrigation to become fertile, this region is the counterpart. The stations on the desert are of no special interest, but as a matter of record may be named as follows: — Ombey, Matlin, Terrace, Bovine, Lucin, Gartney, Tecoma, Montello, Ullin, and Loray. The train has been ascending the grade, and from Kelton, with an altitude of 4,222 feet, to Toano, with an altitude of 5,975 feet, we have made a net gain of 1,753 feet. The mountains to the south are the Toano Range, where mines have been discovered, and which gave a phenomenal output of ore some fourteen years ago, but concerning which, since that time, little has been heard. The great peak almost directly south, which has been our landmark for the last fifty miles is Pilot Knob, rising to a height of twenty-five hundred feet directly from the plains. This Knob was the beacon of the early emigrant by which he steered his ship of the desert, knowing that near it lay Humboldt Wells, where plenty of water and grass could be obtained for his almost famished stock.

Toano. A little station marking the western verge of the desert. (Population small. Distance from Ogden, 183 miles. Elevation, 5,975 feet.)

CEDAR PASS,

The Divide between the Desert and Humboldt Valley.

Highest Elevation, 6,166 feet.

From Toano the ascent of Cedar Pass is begun. For 22 miles the grade is upward, though not remarkably steep, the road rising only 191 feet. The Cedar Pass Range is comparatively low and extends from north to south, the south fork of the Humboldt River flows through these hills. The Ruby Valley lies to the east, and is sixty miles long by ten wide. The valley is occupied by farmers and is very fertile. There are a number of small lakes in the valley, among which may be mentioned Ruby and Franklin.

Moors. This station occupies the summit of Cedar Pass. Snow sheds and fences, which can be seen here and for some distance beyond, testify to the fact that the elevation is such as to cause protection against the danger of snow blockades. (Population small. Distance from Ogden, 210 miles. Elevation, 6,167 feet.)

Wells. The grade has been a descending one since we left Moors, and the descent will be continued for nearly three hundred miles. The railroad company has adopted the monosyllabic title of Wells for this station, but for nearly half a century this place has borne the popular title of "Humboldt Wells." Here the railroad repair shop and round house are located, and the town consists of these and twenty-five or thirty other buildings, including a hotel. In this vicinity, the emigrants in the old days of overland travel to California, were wont to make their camp and recuperate their stock after the trying ordeal of the desert. The wells from which the place takes its name are very curious, consisting of circular openings in the ground varying in size, being from four to eight feet in diameter, and filled to the brink with water. No bubbles arise on the surface of the water, which trickles off through the grass and sinks into the porous soil. It is said that the wells have been frequently sounded and no bottom found. The water is somewhat brackish. There are about twenty of these pools in the little valley, and their life giving influence can be seen in the abundant growth of grass. Because of these peculiar pools Wells is a station of considerable interest to the tourist. (Population, 243. Distance from Ogden, 219 miles. Elevation, 5,628 feet.)

Valley of the Humboldt. After the journey across the desert, the Valley of the Humboldt presents a most delightful appearance to the eyes of the traveler, who is considerably wearied by the constant view of sand and sagebrush. The valley is eighty miles in length and ten in breadth and is occupied by agriculturists and stock raisers. The river which makes this section of the country fertile rises thirty miles northwest of Wells, and, flowing southwest nearly three hundred miles, empties into Humboldt Lake, which has no outlet. The railroad follows the river closely for two hundred and seventy miles and leaves it at Brown's Station, where one has a fine view of the lake. The railroad follows for the greater part of the way the north side of the river, while the old emigrant trail, parts of which can yet be seen, pursues its course on the opposite side of the stream.

Tulasco, Bishops, Deeth, Halleck, Peko, Osino, are all small side track stations, useful to the residents of the valley and to the railroad, but of no especial interest to the tourist. After passing Peko, the railroad crosses the north fork of the Humboldt River and at Osino a cañon of the same name is entered, and we leave behind us the pleasant valley of the Humboldt.

Elko. This is one of the largest towns on the line since leaving Ogden. It is the county seat of Elko County and is well supplied with churches, schools, business blocks and comfortable residences. It is also the seat of the state university. Elko is an important shipping point for stock and for the output of the Eureka, Tuscarora, White Pine and Cape mines, all being within a radius of from twenty-five to one hundred miles. Beyond Elko some ten miles the South Fork of the Humboldt joins the river on the south, watering along its course an excellent grazing country. (Population, 752. Distance from Ogden, 275 miles. Elevation, 5,065 feet.)

Carlin. Between Elko and Carlin is the small station of Moleen. Some hay meadows intervene and the road passes through Five Mile Cañon, where the tourist will behold some rugged scenery. The railroad shops of the Humboldt division of the road are at Carlin. Gold and silver mines within a radius of twenty miles are tributary to the town. (Population, 394. Distance from Ogden, 298 miles. Elevation, 4,897 feet.)

> **THE PALISADES OF THE HUMBOLDT.**
>
> Height of Walls, 1,000 feet.
>
> Objects of Interest, Red Cliff and Devil's Peak.

Twelve Mile Cañon. The road penetrates the range of mountains (which trends from north to south) by way of this cañon. The walls rise on either side in rugged grandeur, attaining in places a height of a thousand feet. From the peculiar stratification of the rocks resembling that of the famous rockwalls of the Hudson, this cañon has been called the Palisades of the Humboldt. Red Cliff is a striking promontory in the midst of the cañon, stained with rubescent colors and rising above the track for more than five hundred feet.

Palisade. This little town nestles in the heart of Twelve Mile Cañon, and is the junction point of the Eureka and Palisade Railroad with the Southern Pacific. The former road is a narrow gauge and was built mainly to convey ore and bullion to the great trunk line. Eureka, its terminus, is a mining town of about six thousand populat on, engaged principally in mining. Here are stamp mills and smelters handling fifty tons of ore daily. Palisade is the site of the machine shops of the Eureka and Palisade Railroad and is also a great shipping point. Beyond Palisade Station is Devil's Peak, an isolated projection on the south side of the river, rising from the water to the height of three hundred feet. (Population, 252. Distance from Ogden, 308 miles. Elevation, 4,840 feet.)

Cluro. A small station which stands at the lower entrance of Twelve Mile Cañon, and is worthy of mention for this fact.

Gravelly Ford. This place is entitled to mention because of its historic interest. It was here that the old California trail crossed the river. The "Ford" was often the scene of Indian raids, and the hardy pioneers and the aborigines more than once tried conclusions here, and the blood of both the white and the red man often stained the flow of the Humboldt.

Beowawe. At this point the Humboldt forces its way through the Red Range of mountains forming a natural "gate," which is the significance of the name Beowawe in the Indian tongue. Beyond the station the road passes through bottom lands covered with a thick growth of shrubbery, the willow predominating. To the south eight or ten miles lies Hot Springs Valley, taking its title from the hot springs which are found there in great number. These springs are intermittent in their flow, resembling in this characteristic, though in a lesser degree, the geysers of the Yellowstone. Beowawe is a station of no very great commercial importance, but possesses interest because of the peculiar features of the surrounding country. (Population small. Distance from Ogden, 326 miles. Elevation, 4,695 feet.)

The Valley Region. To the north and south of the Humboldt and nearly opposite Argenta, are several valleys; among the most important is Paradise Valley—to the north—sixty miles long by ten miles wide, and settled by prosperous ranchmen. Eden Valley, also to the north, is twenty miles long by five miles broad, and thickly settled. Reese River Valley, is to the south, of variable width, not wider than ten miles, and about seventy-five miles in length. The Reese River possesses the peculiarity of sinking into the sand before it reaches the Humboldt, and only in times of great abundance of water does it flow beyond the point of its subsidence.

Battle Mountain. Important as a shipping station for the mining regions in the hills to the north and south ; also the junction of the Nevada Central Railroad with the Southern Pacific. This is a narrow gauge, and its southern terminus is Austin, ninety-three miles distant from Battle Mountain, with a population of three thousand. The Nevada Central penetrates a rich mining district, and not less than twenty camps contribute to its prosperity. Battle Mountain takes its name from the range of mountains to the north of the Humboldt, between the Reese River and Owyhee ranges. (Population, 522. Distance from Ogden, 359 miles. Elevation, 4,511 feet.)

Golconda. A station for the shipment of ores supplied by adjacent gold and silver mines. The Golconda mine is the nearest, being three miles to the south. (Population, 335. Distance from Ogden, 402 miles. Elevation, 4,392 feet.)

Winnemucca. County seat of Humboldt County, and the end of the Humboldt and Truckee divisions of the line. Here are located the shops of the railroad company, which give steady employment to a considerable number of men. The town derives its name from a noted Indian Chief who made his home in this

AT THE GOLDEN GATE.

region. (Population, 2,000. Distance from Ogden, 419 miles. Elevation, 4,333 feet.)

The Nevada Desert. We have now fairly entered upon the Nevada Desert, which we shall travel over to the westward until Wadsworth is reached, a distance of 138 miles. This stretch of country is the most desolate and the most uninteresting of any of the deserts crossed on the transcontinental journey. It is characterized by an almost total absence of vegetation of any kind, and by a remarkable distribution of scoria, the remains of extinct volcanic action. These deposits of black lava are scattered over a grayish expanse of sand, and are of a general cubical form, varying in size from that of a pea to that of a good-sized house.

> **HUMBOLDT.**
> An Oasis in the Desert.
> The Effect of Irrigation.
> Distance from Ogden, 459 miles.
> Population, 32.
> Elevation, 4,236 feet.

As the train stops at Humboldt, the passengers are surprised to see a beautiful little park filled with thrifty trees and carpeted with luxuriant greensward. This oasis in the desert is the result of irrigation, and the fountain of cold, clear water that throws its rainbow tinted spray into the air, tells the story as to how this magical transformation has been brought about. The charm of contrast is complete, and taking all things into consideration, I know of no place to be met with on the trip across the continent that the tourist will regard with more pleasure than the unexpected vision of this emerald of the desert. Star Peak, the highest mountain in the Humboldt Range, crowned with perpetual snow, can be seen only seven miles distant to the northeast, and it is a pleasure to learn that the desert gives way to the Lanson Meadows five miles to the northwest, from which large crops of hay are cut.

Rye Patch. A small station, which derives its name from the fact that wild rye grows here in great quantities. There is in operation here a ten-stamp mill which is supplied with ore from the Eldorado and Rye Patch mining districts lying to the east within a radius of fifteen miles. (Population, 65. Distance from Ogden, 470 miles. Elevation, 4,257 feet.)

Oreana. A small station of no especial interest. A smelter is located here, and the widened expanse of the river at this point is owing to the fact that a dam has been thrown across it to secure water power. The railroad crosses the Humboldt five miles west of Oreana. (Population, 55. Distance from Ogden, 480 miles. Elevation, 4,181 feet.)

Browns. At Browns station the tourist has a good view of Humboldt Lake, as the road approaches it closely. The town itself is of minor importance. (Population, 25. Distance from Ogden, 508 miles. Elevation, 3,929 feet.)

Mirage. Side track station, deriving its name from the phenomenon peculiar to the desert, which has allured many an early emigrant to destruction through its deceptive influences. The green trees, the lake of bright water in which can be seen the reflection of surrounding objects, which the mirage presents to view, are only optical illusions, and those who left the beaten track to seek the refreshment apparently at hand, frequently paid the penalty of their rashness with their lives. (Population, small. Distance from Ogden, 520 miles. Elevation, 4,247 feet.)

THE LAKE REGION.
Facts Concerning Interesting Bodies of Water.

A glance at the map of Nevada will reveal the fact that we have now reached what may very appropriately be called the lake region. These lakes have not the clear, sweet water which one generally associates with the term; but on the contrary are brackish, and hold great quantities of alkali and chloride of sodium in solution. The most important of these lakes are:

Humboldt Lake. This sheet of water takes its name from the river which flows into, or rather through it; the fact being that the waters of the river are collected in this basin, and are then conducted further west into Carson Sink— or Lake. All the drainage carried in the channel of the Humboldt River, in its course of three hundred and fifty miles, is concentrated here; the surplus, as has been said, passing south into Carson Lake which has no outlet. Humboldt Lake is thirty-five miles long by ten miles wide.

Carson Lake. This lake, which receives the waters of the Humboldt River, through Carson Sink, is due south from Humboldt Lake, and has no outlet. The map shows two distinct bodies of water, namely: Carson Sink and Carson Lake; but during the prevalence of rain both are united, and cover a large extent of country. Carson Lake proper, is twenty miles long by ten wide.

Mud Lake is situated north of Granite Point, some fifty miles. The famous "Black Rock" stands at the head of Mud Lake. This promontory is eighteen hundred feet in height, and a strong feature in the landscape. The name of this lake is especially descriptive of its peculiar characteristics, especially during the summer when the water is low and muddy. It has no outlet, and at its season of greatest enlargement is fifty miles long by twenty broad.

Winnemucca Lake is of small extent, being about fifteen miles long by ten wide; it has connection with Pyramid Lake, which lies a short distance to the eastward

Pyramid Lake is made the receptacle of the waters of the Truckee River, the outlet of Lake Tahoe, and is about twice the size of Winnemucca Lake, being thirty miles long by twenty broad.

Walker's Lake has no outlet. It is fifty miles long by twenty wide, and lies about a hundred miles to the south of Mirage.

Hot Springs. A small station, taking its name from the springs which send up the steam from their heated waters on the right of the track. (Population, 42. Distance from Ogden, 535 miles. Elevation, 4,072 feet.)

Desert. This is the last station in the Nevada Desert, marking its western boundary. From here the grade is an ascending one, and when Wadsworth is reached, nine miles beyond, the desert will have been left entirely. (Population small. Distance from Ogden, 546 miles. Elevation, 4,018 feet.)

Wadsworth. The tourist finds a pleasant greeting at Wadsworth, for on arriving at the station he sees a beautiful little park, neatly enclosed and ornamented with a carefully kept lawn and handsome shade trees. The park is not so extensive as that at Humboldt, but is none the less a delight after the long journey across the desert. The town is situated on the eastern bank of the Truckee River, and is prosperous and well built. Here are located the railroad shops for this division of the railroad, and considerable freight business is transacted with the mining camps situated to the south. The Truckee River has its source in lakes Tahoe and Donner, and is a pure and sparkling stream. Six miles south are the Pine Grove Copper Mines, while ten miles south are the Desert Gold Mines tributary to Wadsworth. (Population, 661. Distance from Ogden, 555 miles. Elevation, 4,085 feet).

> **RENO.**
> Junctional Point.
> Distance from Ogden, 589 miles.
> Population, 4,302
> Altitude, 4,497 feet.

In addition to being the county-seat of Washoe County, Reno is a thriving business centre. It possesses all the modern improvements, including electric lights. Its business blocks are well built and its public buildings creditable to the city. The town was named after General Reno, who lost his life in the battle of South Mountain. This is the junctional point for the Nevada & California Railroad, a narrow gauge, the present terminus for which is Moran. Here also the tourist can take the Virginia & Truckee Railroad for Carson City, Virginia City and points to the north and south. Condensing the statement of connections, they are as follows : Virginia & Truckee Railroad for Carson, Virginia and Mound House, connecting there with Carson & Colorado Railroad for Hawthorne (stages for Aurora and Bodie), and for Belleville, Candelaria and Keeler; Nevada & California Railroad for Moran (stages for Millford, Susanville, Quincy, Fort Bidwell, Cal., etc. Stages can also be taken to Eagleville, Alturas, Cedarville and Lake View or Davis Creek. Reno possesses a lively interest to the traveler, as it is the junction point to the world famed Comstock Mines.

Climbing the Sierra Nevada Range. After leaving Reno the grades grow steeper, and the traveler prepares himself for the grand and striking scenery which he will have the pleasure of beholding until the passage of the Sierra Nevada Mountains has been made. For fifty miles the ascent continues until Summit Station is reached, the highest point attained by the Southern Pacific Railroad on its transcontinental line from Ogden to San Francisco. From Reno the road follows up the course of the Truckee River, and soon enters Truckee Cañon. The course of the river is tortuous and the road quickly changes sides, giving varied and interesting views of towering rocks, foaming waters and pine clad mountains. In quick succession the following small stations are passed :

Verdi, Essex, Mystic, Floriston, Boca, Prosser Creek, Proctor and Winsted. The country between Verdi and Proctor seems pretty well given up to the production of lumber, great quantities of ties, logs and boards being piled beside the track. The river is used as a facile means of transporting these products of the forest. Ice store houses also abound here.

Truckee. Roofed like an alpine village to shed the deep, moist snows of winter, Truckee stands at the eastern base of the Sierra Nevada Mountains. The town is well built and extends mainly along the north bank of the Truckee River. Lumber is the leading industry, and where the town now stands, once stood a dense forest. It is estimated that the Truckee Basin will supply at least 4,000,000,000 feet of lumber, or enough to keep the saw mills going at their present rate for a hundred years. The machine shops and round house for this division of the railroad are located here. Truckee is the shipping point for Donner Lake and the towns of the Sierra Valley. Stages can be taken for Lake Tahoe (fourteen miles), Donner Lake (two miles), and Webber Lake; also for Sierraville, Sierra City and Plumas Eureka Mine. (Population, 1,500. Distance from Ogden, 623 miles. Elevation 5,819 feet.)

"There is a grandeur and enchantment at all times in the scenery which environs the lakes of this region and never-ending means of pleasure and exhilaration on their waters; and the panorama of mountain and valley, meadow-land and woodland, sunshine and cloud, as viewed from Tahoe City is spacious, inspiriting and impressive. This view is an unspeakably fine one; within the magnificent frame of the Tahoe range is Lake Tahoe, sometimes tranquil, sometimes turbulent but always lovely. The summer sunsets on Lake Tahoe are remarkable for their great beauty and wealth of coloring and are grander than those mirrored on Lakes Como and Maggiore. No painter would ever dare to put upon canvas the variegated colors of Tahoe's waters in a summer sunset. It would appear such an exaggeration that he would lose caste among those who demand that the artist's pencil shall be true to nature. None but those who have witnessed the scene would be persuaded of its reality. Such beauty could not be were it not for the highly reflective qualities of the pure translucent waters which serve as a polished mirror of French plate glass." Such is the glowing language of a much traveled author, whose words, though eloquent, fail of depicting the entrancing loveliness of the scenes which one can here behold. But it is no reflection upon the descriptive powers of any writer to say that he has fallen short of the reality. Surely if these scenes are beyond the powers of the artist, no discredit can follow when the writer's pen fails to attain to the full measure of their grandeur and beauty.

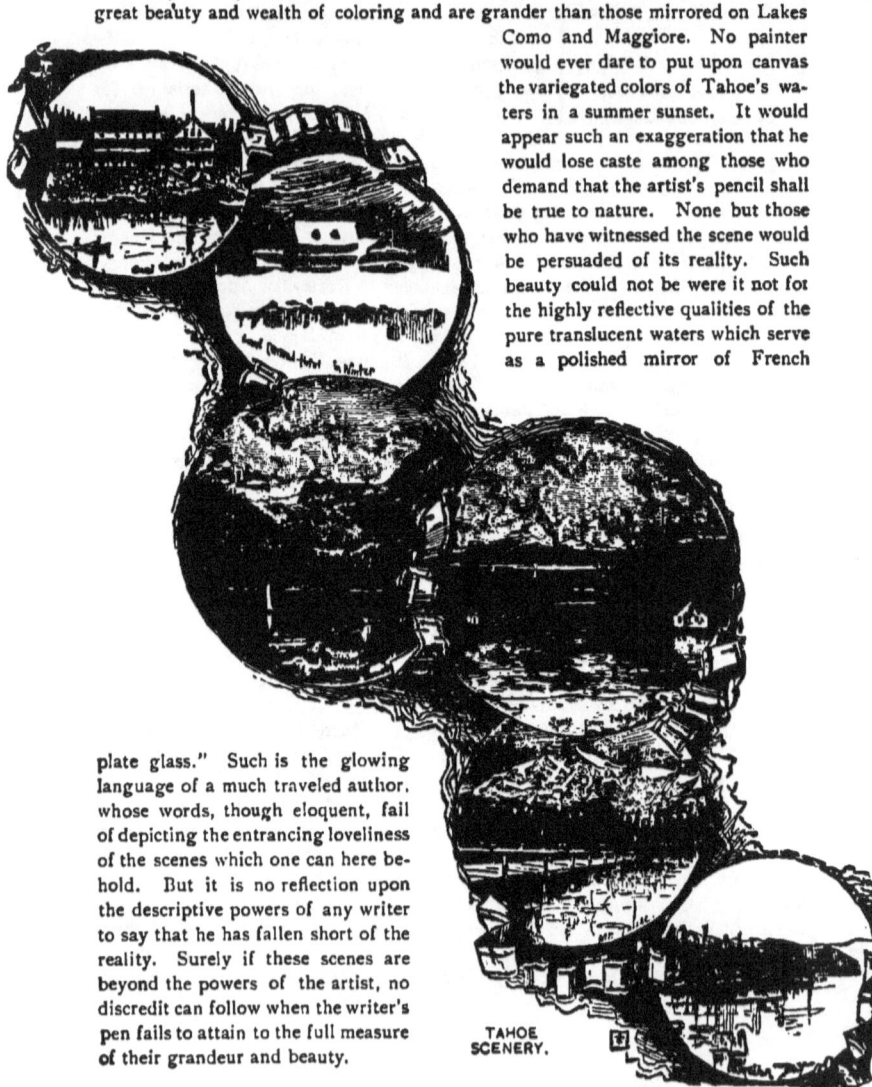

TAHOE SCENERY.

LAKE TAHOE,

The "Gem of the Mountains."

Distance from Truckee, 14 miles.

Length of Lake, 22 miles.

Breadth of Lake 10 miles.

Depth, 1,700 feet.

Lake Tahoe, one of the most beautiful mountain lakes in the world, lies in the heart of the Sierras, 6,216 feet above the sea, while mountain peaks surround it, rising to an additional height of from two to four thousand feet. It is 22 miles in length, 10 miles in breadth and from 100 to 1,700 feet in depth. Its waters are famous the world over for their crystal purity, and their transparency is so absolute that the fish, which abound in great numbers, can be seen distinctly as they swim beneath you, at a distance of eighty feet. On its lovely shores are situated some of the most delightful summer resorts. The mid-summer air is cool and invigorating, the hunting and fishing excellent and the landscape picturesque and a never-ending delight to the eye. The ride by stage from Truckee to the lake, is a most charming experience, and is thus described by Mr. N. H. Chittenden, a traveler of some distinction: "It was a glorious morning, bright and cool, a rain having fallen the previous evening, tempering the dry mountain air, fragrant with the sweet odor of the pines, to a delicious, exhilarating freshness, and also effectually laying the dust. It is a magnificent drive, following up the dashing Truckee, a fitting outlet for the world's crowning gem of mountain lakes. From thirty to fifty feet in width, clear as crystal pure and cold, it courses swiftly down the mountains, frequently a foaming rapid, but interrupted in its headlong descent by several dams. The valley is from three-quarters to a mile across, the mountains generally not precipitous or very high, though presenting several bold, towering granite cliffs and peaks from five hundred to one thousand and eight hundred feet above the river. The most prominent of these, from their resemblance to the human face, are known as the 'Old Woman' and 'Old Man' of the mountains, and the 'Duke of Wellington.' Thick forests of red, yellow and sugar pine, fir and cedar, extend the whole way, except where cleared by the lumbermen. The great saw mill companies are annually cutting millions of feet of the choicest trees, having already advanced about eight miles up the river and back three or four miles therefrom. The lumber flumes extend from the great mills at Truckee to the farthest camps, and the sides of the mountains are grooved with log chutes. Down the former are run vast quantities of wood and timber, while down the latter immense logs are shot, with the velocity of thunderbolts, into the river. At the Eight-Mile Crossing, a five-foot monster plunged in as we passed, striking a forerunner fairly endwise, with terrific force, and the noise of distant thunder. Horse railways and long ox teams are also employed in hauling out the logs from over the summit of the mountains."

The tour of the lake is made by an excursion steamer which is taken at Tahoe City. The surroundings of the lake are picturesque in the extreme. Beginning at the right, the coronet of mountains, which surrounds the lake, may be named as follows: the Rubicon Peaks, 9,287 feet above the sea; Mount Tallac, 9,715 feet in height; Mount Ralston, 9,140 feet; Pyramid Peak, 10,052 feet; Job's Peak, 10,637 feet; Geneva Peak, 9,135, and the summits of the Tahoe Range. Down the steep, forest-covered sides of these mountains swiftly descend numerous beautiful streams, Ward's Creek, Blackwood's, McKinney's, Phipp's, Meek's Bay, Lonely Gulch, Cascade Falls, Cascade Lake, Taylor, Little Truckee River, Big Truckee River, Jim Small's Creek, Sevory Cove Creek, Glenbrook, Secret Harbor, Big, Griffin's, Cornelian Bay and Gordon's Creek being the most important.

DONNER AND WEBBER LAKES.

The shores of Lake Tahoe are indented with beautiful bays, Crystal, Cornelian, Meeks and Emerald, the latter being the largest and most frequented. It is about eighteen miles from Tahoe City, three miles long, and about half a mile in width. Ben Holladay built a summer residence here, which his family occupied until it was burned in 1879.

Capt. Dick, an eccentric old English sailor, chose this wild mountain retreat for his home, built a cabin, and chiseled out a tomb in the solid rock, on the lonely rock-bound island near the entrance. Falling overboard, while intoxicated, Lake Tahoe, which it is said, never gives up its dead, became his last resting-place, instead of the grave he had prepared.

The shores of the lake are dotted with summer residences and pleasure resort villages. Among the latter may be mentioned Tahoe City, Glenbrook, Tallac, Rowlands and McKinneys. Glenbrook is a very pretty village and is the business centre for Lake Tahoe. The thousand and one attractions of this lovely lake can obtain but little justice in so brief a description as can be given here; indeed, the most elaborate description would fall far short of the reality, and only he who has had the extreme good fortune to visit the spot can form an adequate idea of its charms.

DONNER, WEBBER,
AND
INDEPENDENCE
LAKES.
Waters of Crystal Whiteness.

Donner Lake. Made memorable by the terrible fate of the Donner party, thirty-four of whom died of starvation on its shores in the year 1846, and taking its name from the leader of this unfortunate company, Donner Lake commands especial attention for its historical associations. Its beauty gives it a leading position among the lakes of the Sierras and has been made familiar through the well known paintings, by Bierstadt. Only three miles from Truckee, it is easy of access. It is about three miles long, one and a half miles wide, and two hundred and fifty feet deep. Its shores are gravelly and the lake is surrounded by great forests of pine, fir and tamarack.

Webber Lake, a perfect gem, lies in the Sierra Nevadas, about twenty-six miles from Truckee, at an altitude of 6,925 feet above the sea level. It is circular in shape ; its waters crystal white, and with a depth of eighty-four feet. It is considered one of the finest fishing grounds in California, the trout being large and numerous, gamey and delicious. About three-quarters of a mile away from the lake are the falls, having a descent of 105 feet.

Independence Lake, sixteen miles from Truckee, and ten miles from Webber, is another one of those beautiful gems. It is two and one-half miles long and three-quarters of a mile wide. Its waters are alive with trout.

Climbing the Sierra Nevada Range. The ascent of the Sierras begins at Truckee. In order to protect travelers from delay in inclement weather, the railroad company have constructed an almost uninterrupted line of snow sheds for forty miles. These sheds interrupt the view, but they serve an eminently practical purpose and are necessary for winter travel. Through the loopholes cut in the sides of the sheds the tourist catches tantalizing glimpses of magnificent scenery. Donner Lake can be seen below us, gleaming like a diamond in its granite setting, while a panorama of pine-clad hills and splintered mountain pinnacles is spread before us. Plunging onward through the snow sheds, the two great engines drag the train upward, while below can be seen the winding roadway we are ascending. Rumbling through a tunnel the train comes to a halt on the highest railroad point in the Sierras.

SUMMIT.

The Highest Railroad Point in the Sierra Nevadas.

Elevation, 7,017 feet.

Distance from Ogden, 638 miles.

Appropriately named, this station is the summit of our railroad ascent. For many years it held the pre-eminence as the highest railroad point in North America, and it still deserves renown as the first to lay claim to so lofty an estate. This is the "divide" from which flow various streams through devious courses to empty at last at widely divergent points into the great Sacramento. Among these streams are the Bear, the American and the South Yuba Rivers. The scenery around Summit is of the grandest description. The mountains tower above us to an altitude of ten thousand feet. Lakes lie below us and waterfalls glimmer down the sides of distant precipices. Here the sportsman can find ample scope for enjoyment. Bear and deer and a vast variety of game haunt the wooded fastnesses and the streams abound in trout. The east-bound tourist who wishes to visit Lakes Tahoe and Donner can take the stage at Summit, and, after enjoying the delights of the mountain drive and an unobstructed view of the scenery, together with a satisfying visit to the lakes, can again resume his journey by taking the cars at Truckee, thus avoiding the up grade return to Summit.

Cascade. Six miles beyond Summit we pass Cascade, crossing a branch of the Yuba River. To the westward lies Summit Valley, a charming spot for a summer resting place. It is well watered and abounds in luxuriant meadows, which are utilized by stock and dairy men, who have found here an ideal spot for their purposes. Cascade is a growing shipping point for cattle and their products. (Population, 28. Distance from Ogden, 644 miles. Elevation, 6,538 feet.)

Soda Springs. Many large soda springs give their name to this side track. Their waters are pleasant to the taste and medicinal in character. One of the springs has been improved and its waters are bottled for shipment. There are also hot springs in the near vicinity. (Population small. Distance from Ogden, 647 miles. Elevation, 6,749 feet.)

Emigrant Gap. Here we catch the last sight of the old emigrant wagon road, which we have seen from time to time for the last two hundred and fifty miles. (Population, 20. Distance from Ogden, 659 miles. Elevation, 5,221 feet.)

Blue Cañon, Shady Run, Towles, and Alta, are small stations which we pass in rapid succession.

Dutch Flat. Population, 500. (Distance from Ogden, 675 miles. Elevation, 3,595 feet.)

Historic Ground. To the "men of '49" the names of Alta and Dutch Flat call up many memories of stirring times. The stages still run from Dutch Flat to "You Bet" and "Red Dog," where mines are still worked; but the palmy days made historic by the achievements of the "John Oakhursts," "Sandy McGees," and "Hank Monks" have passed away. A glimpse can be caught of a scenic attraction of paramount interest as the train passes Shady Run. This is the famous American Cañon, with walls two thousand feet high, and of such wonderful perpendicularity that the American River, which flows between them, has never been ascended for a distance of two miles—the extent of the cañon.

CAPE HORN.

A Scenic Wonder.

There are few mountain passes more famous than that known to the world as "Cape Horn." The approach to it is picturesque. The north fork of the American River is seen raging and foaming in its rocky bed, fifteen hundred feet below and parallel with the track. A little further on we see the north fork of the North River leaping in snowy cascades down the mountain side. The train rolls on and soon is clinging to the side of a mountain wall, which climbs to the clouds above it and drops to the waters beneath; a hand thrust from the window of the car could drop a stone straight as the plummet falls, into the chasm, two thousand five hundred feet below. We are rounding Cape Horn! The road having been carved from the solid rock, the workmen, when building the same were suspended from the cliff above by means of ropes until they had blasted sufficient to gain a foothold. A beautiful valley lies beneath us to the left, and across this vale on the opposite side can be seen the line of road on which we shall soon appear. The descent now begins, and Rice's Ravine is crossed, the trestle bridge being 878 feet in length and 113 feet in height. The narrow gauge railroad, which we see beneath us, is the line from Colfax to Nevada City. From the trestle we pass to an embankment, and from the embankment to the solid roadway on the side of the bluff. We have followed the curving road until now we are opposite the tremendous precipice, from whose fearful height we have but just descended.

Colfax. Named after the statesman, Schuyler Colfax, a steadfast friend to the Southern Pacific Railroad during the early days of its existence. This town is thriving and prosperous. Fruit raising has taken the place of the original industry of mining, and the financial results appear to be eminently satisfactory. There is a large and handsome depot erected at this place, it being the distributing point for Grass Valley, Nevada City, and a large area of agricultural and mining country. The trains of the Nevada County Railroad (narrow gauge) run to and from this depot. (Population, 400. Distance from Ogden, 689 miles. Elevation, 2,422 feet.)

Auburn. The approach to Auburn is made through a rugged country, a tunnel seven hundred feet in length being passed just before reaching Clipper Gap — beyond this can be seen the famous gold fields, now abandoned. The town of Auburn is embowered with fruit trees, is well-built and prosperous. Many of the residents of San Francisco and Sacramento spend a part of their summers at this mountain town. Fruit raising has usurped the place of mining among these foot hills of the western slope — vineyards, orchards and vegetable gardens, are now seen on all sides. This condition of things exists all along the slope, and for a distance of twenty miles we pass through California's semi tropical fruit belt. The quarrying of stone and stock raising are also important industries. (Population, 1,700. Distance from Ogden, 707 miles. Elevation, 1,360 feet.)

Newcastle. Is situated in the midst of a rich farming region, and is an important shipping point for all California fruits. Here are also a number of extensive canning and fruit drying establishments, with unlimited capacity. The early citrous fruits are grown and shipped from this point. (Population, 350. Elevation, 956. Distance from Ogden, 712 miles.)

Rocklin. This little town lies at the base of the foothills, and is famed for the excellent quality of the granite found in its quarries. The round house and machine shops of the railroad company located here are built of this material.

CAPITOL BUILDING, SACRAMENTO.

The State House at Sacramento is also erected of Rocklin granite. (Population, 800. Distance from Ogden, 721 miles. Elevation, 249 feet.)

Junction. This station is the junction point for the east side of the great Sacramento Valley and Portland, Oregon ; it is here the branch of the Southern Pacific Railroad intercepts the main Transcontinental Line. (Population, 250. Distance from Ogden, 725 miles. Elevation, 163 feet.)

The Plains Region. A glance from the car window, or a reference to the elevation of Junction Station, given in the paragraph above, will show the tourist that the region of mountains and foothills lies behind him, and that the fertile plains of California have been reached. Broad expanses of gently rolling country greet the eye, dotted here and there with the round-topped, dark-foliaged live oaks, which form strikingly characteristic features in the landscape. Here and beyond in the Sacramento Valley are the great wheat fields of the State, famous in the past for their enormous yield and the magnificent scale upon which the raising of this cereal is carried on. Now, however, fruit raising is gradually usurping this territory, and orchards and vineyards are frequently seen.

American River Bridge. This bridge spans the current of the American River, and Sacramento is only three miles distant. (Distance from Ogden, 740 miles. Elevation, 49 feet.)

SACRAMENTO,
California's Capital.

Population, 32,000.

Elevation, 30 feet.

Distance from Ogden, 743 miles.

As is the almost universal rule in the case of large cities one gets a very unsatisfactory view of the town from the railroad station. Several days can be pleasantly and profitably spent by the tourist in Sacramento. It is handsomely built, and its shaded streets and flower ornamented yards present an exceedingly attractive appearance. It has a complete system of street railways, including a recently established and successful line of electric cars Being the capital of California, the county seat of Sacramento County, and the second commercial city in the State, it has a most prosperous present and promising future. More trains arrive and depart each day than in any other town or city in the State. Sacramento, being the geographical centre, it is the great distributing point for California. Three-fourths of all the fruits shipped from this State each year are shipped from this point. It is at this place all the principal buyers and shippers locate for the purchase of fruits and vegetables. The Central Pacific Railroad shops (which employ from 2,000 to 3,000 men constantly, covering an area of twenty-five acres of land), the largest cannery and packing houses in the State, a woolen mill, foundry, machine shops, etc., are located in Sacramento. For a manufacturing town, the location of Sacramento City cannot be excelled. It is ninety miles from San Francisco, with which it is connected by six daily trains, and by river steamers. Many of its wholesale houses rival those at San Francisco in the amount of business transacted. It has fine wide streets lined with shade trees, many substantial business blocks, elegant residences, and good hotels. The State Capitol, State Printing Office, State Agricultural Exposition Building (the largest west of the Missouri river), a Free Library, the largest Art Gallery (with one or two exceptions) in the United States, an Old Ladies' Home (where old ladies have the same care and attention, if not better, than they would have in their own homes), are located in Sacramento City. The two latter were donated to the city by that most estimable and philanthropic of ladies, Mrs. E. B. Crocker. In fact, Sacramento is the great metropolis of the Sacramento valley.

The first railroad in California, extending from Sacramento into El Dorado County, was formally opened on February 22d, 1856 Work on the Central Pacific Railroad was inaugurated at Sacramento, January 8th, 1863, and the last spike was driven May 10th, 1869. Sacramento is on the line of the California & Oregon, Western Pacific, Central Pacific, California Pacific, and Sacramento & Placerville Railroads. All these roads are of the Southern Pacific System. The Company's principal hospital, is also located in this city. A line of steamboats runs to San Francisco on the Sacramento River and the bay, and another as far up the same stream as Red Bluff. The Sacramento River is spanned opposite the city by a railroad and wagon bridge, connecting it with the town of Washington, Yolo County; and the American River is bridged on the line of Twelfth street, and also by a railroad bridge a short distance above. All the bridges in the county and all roads are free. The Capital of California was permanently located at Sacramento, February 25th, 1854, and in 1869 the present Capitol Building was completed, at a cost of about $3,000,000. The building is the finest in the state. In the Capital Park are also the exposition pavilion of the State Agricultural Society, and the State Printing Office, in which are printed, in addition to the usual work for the State, the text-books for use in the public schools. The State Agricultural Society has also an extensive park for the exhibition of stock, and one of the finest race tracks in the world. The State fairs are annually held in September. The Masons and Odd Fellows have each imposing temples, in which their lodge rooms are located. The United States Government has purchased a site for a Post Office Building, to be erected immediately, for which an appropriation of $100,000 has been made. The County Court House (formerly used for a State Capitol) cost $200,000; and a brick and iron Hall of Records has recently been completed at a cost of $50,000. The County Hospital built on the pavilion plan, can accommodate one hundred and seventy-five patients, and cost $75,000. The State Library contains some sixty thousand volumes; the Free Public Library, of twelve thousand volumes, with the two story building in which it is contained, is the property of the City, and is maintained by a City tax. The Order of Odd-Fellows maintain a library of about eight thousand volumes. The Crocker Art Gallery is also the property of the City. It is a brick and iron building, three stories high, and in it are contained some of the finest paintings and statuary, together with an extensive cabinet of minerals, the property of the State.

Webster. Leaving Sacramento, and crossing the Sacramento River on a bridge 600 feet in length, the train passes through Webster, which is a suburb of the city. Beyond we cross a belt of swampy country known locally as "The Tuiles." The track is elevated above the danger of floods by means of embankments and a trestle bridge.

Davis. This place is the junction with the main line of a branch passing through the west side of the Sacramento Valley to Tehama, the country round about being rich and fertile, and capable of producing an unlimited amount of fruit, cereals and vegetables. Distance from Ogden, 736 miles.

Fremont, Dixon, Batavia are soon passed, when we arrive at

> **ELMIRA.**
>
> Junction Point
> to
> Vaca and Capay
> Valleys.

At this point the tourist will do well to take the side trip through the great Vaca and Capay Valleys. These valleys supply all the earliest fruits and vegetables. The soil is of surprising fertility, yielding bountifully of every crop with no necessity for irrigation. The climate is superb. it being a continual Indian summer the entire year. The health of the inhabitants, their industry, wealth and prosperity, have all tended to make this place the most desirable for settlement. Semi-tropical and citrus fruits grow luxuriantly, and are of unusual size and lusciousness. These valleys are veritable gardens of Eden, and a continuous panorama of a beautiful and picturesque country. Cannon and Suisun are more or less important stations, but of no especial interest to the tourist. Having passed Suisun the waters of Suisun Bay approach the track, and at high tide ripple against the embankment. For twelve miles this bay is always in close proximity.

Army Point. Distance from Ogden, 797 miles. This is the station for the headquarters of the United States army in California.

Benicia. Situated on the southern slope of the Suscal hills, Benicia extends down to the bank of the Sacramento River. This is the head of navigation for sea-going ships and is a very charmingly situated city. Benicia was at one time the capital of California, but is now a quiet residence town, with a number of large manufacturing interests to maintain its commercial importance. (Population, 3,200 Distance from Ogden, 800 miles. Elevation, 10 feet.)

Crossing the Straits of Carquinez. From Benicia to Port Costa the journey is continued on the Solano, the largest ferry boat in the world. This boat can transport at one time fifty-four loaded freight cars and consequently finds no difficulty in bearing our entire train safely across the straits, a distance of one mile, with an expenditure of little if any more than twenty minutes of time. To most, this experience is a novel one, and the cars are quickly emptied by their occupants, and the tourists gaze delightedly at the broad expanse of waters and inhale gratefully the invigorating saline odors wafted from the neighboring ocean. The cars are run directly on to the boat and when Port Costa is reached the journey by rail is resumed.

Port Costa. Here the sea-going ships can be seen lying close to the wharfs, and the tourist begins to appreciate the fact that his long journey to the Pacific coast is nearly completed. At this point the Southern Pacific's line to Los Angeles branches to the southwest.

Vallejo Junction. The town of Vallejo lies across the straits a distance of two miles. At this junction a branch line runs to Napa and Calistoga, also to Santa Rosa.

Pinole. Another town of wharfs and warehouses.

Sixteenth Street, Oakland. This is the small station for the large city of Oakland. The great Bay of San Francisco lies to our right and beyond can be seen the spires of San Francisco.

Oakland Pier. This marvel of engineering has been constructed for two miles directly out into the bay. At its terminus is an immense building containing waiting rooms and all necessary accommodations for the convenience of the great army of travelers who disembark on the arrival of trains. All the passenger trains for the east, north or south are made up at this depot, and here all incoming passengers leave their trains and are transported on magnificent ferry boats to San Francisco.

APPROACHES TO OAKLAND FERRY

SAN FRANCISCO TO SAN DIEGO.

SAN FRANCISCO. The Great City by The Golden Gate. Population, 400,000.

The first view of San Francisco which the overland tourist obtains from the bow of the ferry boat that bears him from Oakland Pier to the foot of Market street, is most enchanting. A city set on a hill, beautiful for situation, it commands attention and demands the most enthusiastic admiration. Nor does "familiarity breed contempt." The first pleasant impression is confirmed and deepened by every day's experience within the gates of this most hospitable and beautiful city. Fitz Hugh Ludlow, whose early death was a great loss to literature, if one may judge by the early fruitage of a tree too soon cut down by cruel frost, speaks glowing words, and true ones of this city by the sea. He says: " To a traveler paying his first visit, it has the interest of a new planet. It ignores the meteorological laws which govern the rest of the world. There is no snow. There are no summer showers. The tailor recognizes no aphelion or perihelion in his custom; the thin woolen suit made in April, is comfortably worn until April again. Save that in so-called winter rainfalls alternate with spotless intervals of amber weather, and that *soi-disant* summer is an entire amber mass, its unbroken divine days concrete in it, there is no inequality on which to forbid the bans between May and December. In San Francisco there is no work for the scene-shifter of Nature. The wealth of that great dramatist, the year, resulting in the same manner as the poverty of dabblers in private theatricals—a single flat doing service for the entire play. Thus, save for the purposes of notes of hand, the almanac of San Francisco might replace its mutable months and seasons with one great, kindly, constant, sumptuous All the Year 'Round. Out of this benignant sameness what glorious fruits are produced! Fruit enough, metaphorical; for the scientific man or artist who cannot make hay while such a sun shines, from April to November, must be a slothful laborer, indeed. But, fruit also literal; for what joy of vegetation is lacking to the man who, every month in the year, can look through his study window on a green lawn, and have strawberries and cream for his breakfast. Who can sit down to this royal fruit, and at the same time to apricots, peaches, nectarines, blackberries, raspberries, melons, figs, both yellow and purple, early apples and grapes of many kinds."

But aside from the claims of climate, which appealed so strongly to Ludlow, San Francisco has artistic and architectural claims that command respect and admiration, to say nothing of her vast commercial and mercantile interests.

San Francisco has suffered greatly from fire in the past, but has always arisen from its ashes in renewed beauty. A condensed history of these great conflagrations may be of interest:

December 24th, 1849. First great fire. More than $1,000,000 worth of property destroyed.

May 4th, 1850. Second great fire. Three blocks of buildings consumed. Loss, $4,000,000.

June 14th, 1850. Third great fire. Loss, $5,000,000.

September 17th, 1850. Fourth great fire. An extensive area of comparatively inexpensive buildings destroyed. Loss, $500,000.

December 14th, 1850. Fire on Sacramento and Montgomery streets. Loss, $1,000,000. This is not generally classed among the great fires.

May 4th, 1851. Fifth great fire. Eighteen blocks entirely burned, and parts of six others destroyed. The length of the burned district was three-fourths of a mile, and its width half a mile. Loss, $10,000,000 to $12,000,000.

June 22d, 1851. Sixth great fire. Ten blocks and parts of six others destroyed. Loss, $3,000.000.

When the Oakland ferry boat, a most magnificent steamer by the way, enters her pier at the foot of Market street, the traveler will find ample means of conveyance to any hotel. If of an economical turn of mind he can board a cable car, after running the gauntlet of vociferous "cabbies," and for five cents be carried smoothly and quickly to almost any part of the city; or, handing his baggage checks to one of the agents of the United Carriage Company, he can drive to his destination in considerable more "style," and at a moderate expense, the amount being determined by the distance traveled—but extortion need not be feared, as cab fares are regulated by a city ordinance. Once at home in hotel or lodgings—and San Francisco can furnish either of these of the very best character—the traveler can map out excursions in the city and its environs that will pleasantly occupy his time for a fortnight, or which can be crowded into the space of three or four days.

CLIFF HOUSE AND SEAL ROCKS
Novel and Characteristic Attractions.

Everybody has heard of the Cliff House and the Seal Rocks. These attractions are pretty sure to command first attention. The Cliff House may be reached by three routes These are tersely described by Mr. Charles Turrell, in his valuable California notes, as follows: "One of these routes is the old road that begins at the Mission and winds over the hills, affording many attractive views of the city and the bay beyond, the Contra Costa Mountains and Mount Diablo towering in the remote east. This road descends to the Ocean beach, passing near Merced Lake—Laguna de la Merced—the largest lake in the county. From the Ocean Side House to the Cliff House, a distance of some two and a half miles, the road follows the sandy beach. As this road is quite long, and the latter part very heavy, but few follow it. Another route is by Point Lobos avenue, a broad, well macadamized street, commencing at the western end of Geary street and continuing in a straight line to the Ocean beach. This was for many years the fashionable drive for San Franciscans. However, since the Golden Gate Park has been opened, and its serpentine drives to the beach completed, the Point Lobos road has fallen into disuse." This drive is the one we took, and we found it a most charming way. The Haight street cable car for Golden Gate Park took our party to the entrance of the Park, and here a carriage was engaged for the drive to the Cliff House and return; thus economy was subserved and nothing of pleasure lost. The Park, though in a state of transition from wild land to a cultivated Paradise, presented many most charming views. The abundance of natural flowers, the flora new to our unaccustomed eyes, the conservatory abounding in tropical flowers, the shaven lawns, and the artistically arranged trees and shrubbery, were objects of great interest. From Inspiration Point we obtained a fine view of the Pacific Ocean and the Golden Gate. The most characteristic objects of interest at the terminus of this drive, are the Seal Rocks and their curious occupants. The rocks are conical in shape, three in number, and vary in height from twenty to fifty feet. These rocks are the haunts of seals, and it is said that there is never a moment when scores of these curious marine mammals may not be seen basking in the rays of the sun on these rocks, or struggling among themselves for a place thereon. These seals are pro-

VIEWS FROM THE CLIFF HOUSE.

tected by law, and there is, therefore, no great danger of future travelers visiting Seal Rocks only to be disappointed.

SAN FRANCISCO BAY.

A Beautiful Sheet
of Water and
Land - Locked Harbor
of
Inestimable Value.

San Francisco Bay. As a harbor it ranks among the few great seaports of the world. A land-locked sheet of water, some fifty miles long and of varying width. It has the advantage of lying at the central edge of a great area of agricultural land. The shipments through this port are very heavy, giving constant employment to a large fleet of steamers and sailing-vessels. It is also the terminal point of the great transcontinental routes. If the tourist will take a seat on the dummy of either the California Street or Jackson Street cable cars and ride as far as Mason Street, the trip will be amply rewarded. Perhaps the best time to view this magnificent panorama would be in the forenoon. To the left we have the Golden Gate, the wonderfully beauteous entrance to the still more beautiful bay; to the right the sheet of water merges into the distant hills bordering the Santa Clara Valley. Before us lie, in semi-circular form, Mt. Tamalpais, standing on the northern side

BIRD'S-EYE VIEW OF SAN FRANCISCO, CAL.

of the Golden Gate ; Saucelito, San Pablo Bay, the *debouchere* of California's two great rivers—the Sacramento and San Joaquin ; then we have the Contra Costa Mountains and, just beyond, Mount Diablo's graceful peak, while nestling at their base we distinctly trace the towns of Berkeley, Oakland, Alameda, Haywards, and Oakland Pier. The steamers of the ferry lines may be viewed ploughing their rapid way to and from San Francisco. Close to the Pier, Goat Island rises three hundred and forty feet out of the water. It is the most southerly island in the bay, save the Mission Rock, now surrounded by warehouses, etc. East of Goat Island is Alcatraz Island, situated about one mile due east of the Golden Gate, whose entrance it commands. It is one-third of a mile long and one-tenth of a mile wide, irregular in shape and contains about twelve acres, composed mainly of solid rock. A perfect belt of batteries surround the island, mounting several very heavy guns on all sides as well as on the top. On the highest point of the island stands a light-house, whose light can be seen, on a clear night, twelve miles at sea, outside of the Golden Gate. Next in succession is Angel Island, three miles north of San Francisco, the largest and most valuable island in the bay. It contains six hundred acres of excellent land, watered in many places by natural springs. Three fixed batteries, mounting large, heavy guns, are here, besides large barracks, accommodating the garrison. On the bay we see craft of every kind, from the tiny skiff to the monster six-masted ocean steamers. Scows and steamers may be seen in every direction; the propeller, the paddler are all here in busy activity. Fringing the water front is a forest of masts, the black hulls from whence they spring being scarcely visible on account of the long line of the sea-wall and warehouses that intercept the view. In every direction, lying peacefully at anchor, are vessels just arrived or about to depart. Here, too, snugly harbored, are the little yachts of the different clubs—white-winged birds of pleasure.

There are several "squares" in San Francisco, the most noted of which is Portsmouth Square, with an area of 275 by 204 feet 2 inches. Its history is important. On July 8th, 1846, Captain Montgomery, of the United States sloop-of-war Portsmouth, then lying in the bay, at the command of Commodore Sloat, raised the American flag on the plaza of what was then called "Yerba Buena"—now San Francisco. A salute of twenty-one guns from the Portsmouth announced the fact that the United States had taken possession of Northern California. This square was then named Portsmouth Square, and at the same time Montgomery street was named in honor of the Captain.

Telegraph Hill is dear to the hearts of old Californians. In 1849 a signal station was established on this elevation, and the dwellers at the "Bay" were notified of the approach of vessels from sea by means of a well understood system of signals. A tract of 275 feet square on the summit of the hill has recently been purchased by some public spirited citizens and presented to the city for a perpetual park.

Many tourists take interest in the cemeteries of a city ; to such a brief mention of those in San Francisco will be interesting. Most of these "cities of the dead" are best reached via the Geary Street Cable Railway. Laurel Hill Cemetery, near the foot of a solitary hill, called Lone Mountain, presents the finest examples of mausoleum architecture in California. Landscape gardening contributes greatly to the beauty of the scene.

The four principal cemeteries of the city surround Lone Mountain. They are "Laurel Hill," "Calvary," the Roman Catholic burial ground, and the cemeteries of the Masons and the Odd-Fellows.

Woodward's Gardens, with an area of about six acres, filled with attractions of a most varied character, are greatly frequented. Here are museums, conservatories, aviaries, zoological collections, and a great wealth of floral beauty. Half a day can be spent here with both pleasure and profit. The admission fee is only twenty-five cents.

> **THE MISSION DOLORES.**
> Oldest Building in San Francisco.
> Founded Oct. 8, 1776.

The oldest building in San Francisco and the one most noted, considered historically, is the Mission Church, on the corner of Dolores and Sixteenth Streets. Considerable of the original building remains and many of the interior decorations have been, to a certain degree, retained in their pristine state—sufficient to recall the times of the early fathers. The adobe walls are three feet thick, resting on a low foundation of rough stone, not laid in mortar; and the roof is covered with heavy semi-cylindrical tiles. The floor is of earth, except near the altar, and the entire structure rude in character and still used for purposes of worship.

IN SAN FRANCISCO BAY.

Adjoining it is the Mission Cemetery, not used for purposes of interment since 1858. Most of the inscriptions on the tombs are in Spanish. Clustering around the mission are a few adobe buildings, red tiled but dilapidated, yet speak to the thoughtful of five score years and more. It is best reached by taking the Castro Street cable car of the Market Street Railway.

The theatres are numerous and first class, but English theatres are the same in kind the world over, and need no special description. Not so, however, with the Chinese theatre. This is *sui generis*, entirely novel and of remarkable interest. There are two of these theatres in San Francisco, and the histrionic peculiarities of the Celestial drama can here be seen in greater perfection than in any other city in the world, with the exception of those of China. There is no danger in visiting these theatres, as they are as well conducted, in their peculiar Chinese way, as any other place of amusement; but if there is a party, especially if it contains ladies, the escort of a guide should be secured. Through his influence and acquaintance seats can be obtained upon the stage, and a fine view of the wonderful performance obtained. The stage has no scenery. The orchestra occupies the back of the stage, and the most industrious member of it is the man who manipulates the big bronze cymbals and the gongs. This fellow punctuates the dialogue with vigorous blows on his loud resounding instruments, giving to the drama the characteristic of operatic recitative. The other instruments are the Chinese violin and fife. The result is a queer kind of barbaric harmony, but to the English ear there is absolutely no melody. The "property" man sits on the stage in full view of the audience and supplies the actors with such properties as they may need during the action of the play. The actors are masters of their art. They possess great facial mobility, and even through their conventional "make up" one can recognize their histrionic ability. No women are allowed to act in the Chinese dramas, and all female characters are played by men. These actors are exceedingly clever, and in voice and action imitate the weaker sex most admirably. A good female impersonator receives a very large salary from the management. Whenever it is necessary to personate a death upon the stage, the actor lies quietly for a moment, and then calmly rises and walks off. A stick with a tuft of horse hair represents a horse, and a gesture of the leg signifies that the cavorting animal has been mounted. After all, these conventionalities are not much more crude than those of the Shakesperian age. The dramas are historical, and some of them are more extended even than a Wagnerian triology—requiring from three to four weeks to present a single play.

It would be vain for the writer to attempt to give a circumstantial description of the attractions of San Francisco. It would require a volume, and the pen of a Bayard Taylor to do the city justice. As a convenience for strangers, the following list of places of amusement and points of general interest is annexed:

NEW BALDWIN THEATRE—Baldwin Hotel. Market and Powell.

THE ALCAZAR—O'Farrell street, between Stockton and Powell.

BUSH STREET THEATRE—Bush street, above Montgomery.

BIJOU THEATRE—Market street, opposite Grant avenue.

TIVOLI OPERA HOUSE—Eddy street, near Baldwin Hotel. Grand operatic performance every evening. Grand orchestra and chorus. Admission, 25 cents. Extra to reserve.

PANORAMA BUILDING—Southwest corner Eddy and Mason streets. Open daily (Sundays included) from 9 a.m. to 11 p.m. Admission: Adults, 50 cents; Children, 25 cents.

PANORAMA BUILDING—Corner Tenth and Market streets. Open daily (Sundays included) from 9 a.m. to 11 p.m. Admission, 50 and 25 cents.

"ORPHEUM" OPERA HOUSE—O'Farrell street, opposite "Alcazar." Admission, 25 cents. Extra to reserve.

CHINESE THEATRE—Grand Chinese Theatre, 814 Washington street. Performances every evening by full Chinese Company. Admission, 50 cents. Private Boxes, $3.00.

GOLDEN GATE PARK—Contains over 1,000 acres; extends from Baker street to the Pacific Ocean, 3½ miles. Reached by Market Street Cable Railway via Haight, Hayes, or McAllister streets, from ferries; or, Geary Street Cable Road, from corner of Kearney and Geary streets; and via Powell or California Street Cable Roads. Fare, 5 cents.

WOODWARD'S GARDENS—Reached by Valencia Street Division of Market Street Cable Railway. An extensive and beautiful park, filled with trees, flowers, and rare plants, menagerie, botanical garden, aquarium, and museum of curiosities. Performances on Saturdays and Sundays. Admission, 25 cents. Children, 10 cents.

GLIMPSE OF CELESTIAL LIFE IN SAN FRANCISCO

CLIFF HOUSE AND SEAL ROCKS—Point Lobos, 6 miles from City Hall. A magnificent drive over a perfect road leading through Golden Gate Park ; or, can be reached by Market Street Cable Railroad, Haight Street Division, connecting at terminus with trains of Park & Ocean Railroad direct to Ocean Beach, near Cliff House. Distance from Oakland Ferry, about 8 miles ; time, 55 minutes; fare, 10 cents. Also reached by Powell Street Cable Railroad and Ferries, and Cliff House Railroad.

SUTRO HEIGHTS—The private garden of Adolph Sutro, made beautiful beyond description by the gardener and artist, is just back of the Cliff House, but higher up. Open daily from 10 a. m. to 5 p. m.

PRESIDIO RESERVATION—Fronts on the Golden Gate for about two miles. It has several beautiful drives, is owned by the Government, and its barracks have the largest military force on the Pacific Coast. Drive out California Street or take California Street, Jackson Street or Union Street cable cars.

POSTOFFICE—Corner of Washington and Battery Streets. General delivery is open from 7:30 a. m. to 6 p. m. every day, Sundays excepted ; Sundays, from 1 to 2 p. m. Branch postoffice, station "A," Polk and Austin streets ; " B," Eighth and Mission Streets ; "C," Twentieth and Mission Streets ; " D," foot of Market Street, at ferries.

MARKETS for fruit, flowers, fish, game and other produce : California Market, California Street, below Kearney ; Centre Market, Sutter and Grant Avenues. Visit early in morning. Semi-tropical fruits and flowers all the year round.

SAN FRANCISCO STOCK EXCHANGE—Pine street, between Montgomery and Sansome.

MERCHANTS EXCHANGE—California street, between Montgomery and Sansome.

UNITED STATES MINT—Fifth and Mission streets. Visitors admitted from 9 a.m. to 12 noon, except Saturday and Sunday.

CALIFORNIA STATE MINING BUREAU—New Pioneer Building, Fourth street. This institution has the largest and most valuable collection of ores, minerals, fossils, and Indian relics, in the United States.

MISSION DOLORES—Founded 1776 ; 16th and Dolores streets. Reached by Valencia Street Division of Market Street Cable Railway.

ALCATRAZ ISLAND AND ANGEL ISLAND—Permission to visit these may be secured at department headquarters, Phelan Building, Market St., except Sundays. Steamer General McDowell visits them daily.

EASTERN RAILWAY LINES—The offices of all agents of eastern railroads, represented in San Francisco, are on Montgomery, Market and New Montgomery streets ; in close proximity to Palace, Grand and Occidental Hotels.

BANKS—All the leading banks are in block bounded by Montgomery, Sansome, California and Pine streets.

EXPRESS OFFICES—Wells, Fargo & Co., corner Mission and New Montgomery streets, opposite Palace Hotel.

STREET CAR FARES—The fare on all street car lines, both horse and cable, is 5 cents.

HACK FARE—One person, not more than one mile $1 50
        Two or more persons   "   " 2 50
        Four or less, by the hour—first hour............ ... 3 00
        Each subsequent hour 2 00

PAVILION, WOODWARD'S GARDENS.

CABS—One person, not more than one mile	50
Two or more persons, by hour—first hour	1 50
Each subsequent hour	1 00

> **OAKLAND.**
> Beautiful Residence City.
> Population, 55,000.
> Distance from San Francisco, 8 Miles.
> Elevation, 12 Feet.

Oakland. It is to be supposed that the tourist in his stay in San Francisco has not neglected to visit this garden city. The town is beautifully situated on the east shore of the bay, the land sloping gradually down to the waters from the Contra Costa Mountains, which rise back of the city at a distance of a few miles. The foot hills are crowned with the suburban villas of wealthy merchants of Oakland and San Francisco, and from their verandahs can be obtained a most extensive and pleasing view of the bay, San Francisco and the Ocean beyond. Oakland is one of the most beautiful residence cities in the world, and in point of sylvan beauty has few if any rivals. The houses are tastefully built, many of them of the greatest elegance, surrounded by extensive and well kept grounds, embowered in trees and glowing with a lavish wealth of roses. It must not be supposed, however, that Oakland is not also a business town. On the contrary, it possesses large mercantile and manufacturing establishments. Electric lights illuminate the wide and well paved streets; cable and horse car lines are numerous and none of the modern improvements lacking. Schools and churches abound. Oakland is a city of colleges, and numbers among these institutions of higher education the following: The State University School, the Oakland Military School, the Convent of Our Lady of the Sacred Heart, the Oakland Female Seminary, the Female College of the Pacific, and the University of California, at Berkley, four miles distant. Among the large manufacturing establishments may be mentioned the extensive machine shops of the Southern Pacific Company, the Judson Manufacturing Company, the Pacific Iron and Nail Company, besides cotton mills, jute mills, flour mills, and innumerable other institutions, employing a large amount of capital and thousands of men, women and children. One can reach San Francisco from Oakland every fifteen minutes by train and ferry. Oakland is a most charming place, and is the home of an enterprising, hospitable, and intelligent class of people.

Southward Bound. Having spent a most delightful season in San Francisco, the tourist's face is turned southward, and the journey to Los Angeles and San Diego begins. Taking the Oakland ferry, at the foot of Market street, one is borne pleasantly over the waters of the bay and lands at Oakland pier, where he takes the Southern Pacific train for Los Angeles.

Doubling on our Track. From Oakland to Port Costa we follow the same line as that upon which we entered San Francisco, therefore, it is not necessary to make mention of the intervening stations. Passing Port Costa, the line has the Sacramento River on its left, and rolling hills on its right. Beyond the river can be seen the town of Benicia nestling among the coves of the Suscal Hills.

Martinez. A pleasant village among the hills. Fruit trees and vines abound, and the inhabitants of the towns and surrounding country are mainly engaged in horticulture. Martinez is the county seat of Contra Costa County, and is a most quiet and charming place of residence. Citrus fruit, grapes of all varieties, and deciduous fruits flourish without irrigation, and the climate is so mild that semi tropical plants grow out of doors without any special protection. (Population, 1,500. Distance from San Francisco, 35 miles. Elevation, 10 feet.)

ON WHEELS, THROUGH GOLDEN GATE PARK.

Avon, Bay Point and Cornwall are small intermediate stations.

Coal Mines. About six miles south of Cornwall are large coal mines, the tramways for the conveying of the product of these mines pass over our track, and deliver the coal at Pittsburgh Landing on the river, whence it is carried by water to destination. From Martinez to Antioch the road passes through a hill country on our right, with the river to the left. Many deep cuts occur, and numerous small tributaries flow down the gulches, into the river. Up these gulches we catch glimpses of neat farm houses, surrounded by well cultivated fields and orchards. Mount Diablo rises to the south, and reaches an elevation of 3,896 feet. Among the foot hills of this mountain are the mining towns of Stewartville, Empire, Nortonville and Somerville. At Cornwall to our left lies Suisun Bay, and here the San Joaquin and Sacramento Rivers have their junction.

Antioch. A shipping point for coal. The town itself is a mile north on the banks of the San Joaquin River. From this point also large quantities of vegetables, strawberries, fruit, etc., are shipped to San Francisco. (Population, 700. Distance from San Francisco, 55 miles. Elevation, 46 feet.)

Bentwood. Wheat fields begin to appear here, dotted with live oaks. The town is small and supported by agricultural industries. It is situated on the Marsh Grant of 13,000 acres, on which much stock is fed.

Byron. The most attractive thing about this station, to the invalid and the tourist, is, its near proximity to the Byron Hot Springs, situated two miles to the south. The country round about is famous for its production of wheat, alfalfa, fruit and grapes. This being a portion of the great wheat belt. The hot springs have attracted much attention, and a large hotel and bath houses have recently been erected. The springs are varied in their characteristics, being both hot and cold, and possessing in turn the constituents of sulphur, iron, soda and magnesia. There are mud baths, and in fact all varieties of bathing. The temperature of some of the springs is as high as 130° Fahrenheit.

```
BYRON HOT SPRINGS.
    Bathing
       and
 Health Resort.
```

Bethany. Distance from San Francisco, 76 miles.

Tracy. The junction of the old Western Pacific route from San Francisco to Sacramento via Livermore Pass with our line to the south. Tracy is surrounded by broad wheat fields, which extend to the northward beyond the reach of vision. (Population, 200. Distance from San Francisco, 71 miles. Elevation, 64 feet.)

Banta. Small station three miles from Tracy, after passing which we cross the San Joaquin River on a very long draw bridge. (Population, 150. Distance from San Francisco, 74 miles. Elevation, 30 feet.)

Lathrop. Junction of the old Western Pacific and the Sunset Route. This is a regular meal station and here the railroad company have erected a large hotel, in which are also their offices. Lathrop is in the heart of the great San Joaquin wheat belt. (Population, 600. Distance from San Francisco, 83 miles. Elevation, 26 feet.)

The San Joaquin Valley. After crossing the San Joaquin River and turning to the right, our course is up the famous San Joaquin Valley — the great granary of California. Here are five million acres of the best wheat land in the world. A valley two hundred miles long by thirty miles broad, which when vivified by the magic touch of irrigation, produces not only wheat but also almost every thing that can be raised in tropical or temperate zones — wheat, corn, oats, flax, apples, oranges, lemons, figs, nuts, olives — the list is too extended for recapitulation. Properly conserved there is water enough to irrigate the whole valley, and in many places the natural supply of water has been supplemented by that flowing from artesian wells. After passing Lathrop, we rattle through a number of small stations, all of them with large shipping warehouses, speaking eloquently of the generous output of the soil.

Passing through Morano, Ripon, and Salida, small stations, we reach **Modesto.** County seat of Stanislaus County, and a prosperous and pretty town, surrounded by an industrious agricultural people. (Population, 2,500. Distance from San Francisco, 114 miles. Elevation, 91 feet.)

Between Modesto and Merced are the unimportant stations of Ceres, Turlochs, Livingston, and Atwater.

Merced. A well-built town, the county seat of Merced County. Possessed of good public buildings, fine private residences, and surrounded by an exceedingly rich agricultural country, and destined to be a great manufacturing center, Merced has prospered and will continue to prosper. The county has a population of 75,000, nearly all engaged in agricultural pursuits. (Population, 3,000. Distance from San Francisco, 152 miles. Elevation, 171 feet.)

Athlone. Before Athlone is reached we cross the Mariposa River, and after it is passed the Conchilla River. Wheat fields are on every hand. Irrigating ditches abound. Vineyards are frequently to be seen. And Athlone, a quiet little village, sits in the midst of fertile fields. (Population, 50. Distance from San Francisco, 162 miles. Elevation, 210 feet.)

BERENDA.

Junction Point

to the

World's Famous

Yosemite Valley,

Big Trees, etc.

This station is situated at the junction with the main line of the Yosemite extension of the Southern Pacific Railroad, which extends to Raymond, a distance of twenty-one miles to the eastward. From Berenda a good view of the Sierra Nevada Mountains can be had. Among the highest peaks in view are those of Mount Lyell, Mount Tyndal, Mount Goddard and Mount Whitney. These mountains, which exceed 14,000 feet in altitude, impress one deeply with their vast proportions, more especially because we are so near the sea level, being at an elevation of less than three hundred feet. Berenda has an agricultural and grazing country directly tributary to it. (Population, 85. Distance from San Francisco, 178 miles. Elevation, 256 feet.)

Madera. This is a leading shipping point for lumber, which is delivered to this point from the foot-hills by means of a flume fifty-three miles in length. The great work of constructing this flume was completed in 1876, which has been in service ever since. The amount of lumber delivered in this way during the last ten years is something enormous, as may readily be gathered from the fact that last year's delivery amounted to over twenty-two million feet. (Population, 700. Distance from San Francisco, 185 miles. Elevation, 278 feet.)

Fresno. Between Madera and Fresno there is some interesting country. Just after leaving Madera we cross the Fresno River, beyond Sycamore the San Joaquin River, and at Borden, Cottonwood Creek. The sand dunes will attract your attention beyond Sycamore—queer little hills of sand fifteen to twenty-five feet in diameter and three to six feet high. Fresno is the county-seat of Fresno County, and is a most thriving and prosperous city. It has electric lights telephones street railroads, water works, in short, all the modern improvements. Redwood and pine is the material mostly in use for building purposes, and the town possesses many elegant public and private edifices. A great variety of industries are tributary to the town. Fresno County has about 30,000 acres planted to grapes, and shipped last year over five million pounds of raisins. This is but a small part of the product of the county. The shipments of various farm products reached the high figure of one hundred and sixty million pounds of freight. There is an abundant supply of water for irrigation, being

THE PETRIFIED FOREST

brought from the mountains by means of canals having an aggregate length of eleven hundred miles and costing two million dollars. The capacity of these canals for irrigation covers a space of over seven hundred thousand acres, thus making Fresno County one of the richest agricultural regions in the world. Lombardy or the Nile Valley are not richer in possibilities. Many colonies have formed settlements in the vicinity of Fresno. These enterprises, through intelligent and united industry, have proved very successful. With a salubrious climate, fine scenery, fertile land and an industrious people, Fresno has every reason to anticipate a continuance of her phenomenal success. (Population, 12,000. Distance from San Francisco, 206 miles. Elevation, 293 feet.)

Selma. Surrounded by a wheat growing country and supplied with good flouring mills, this town is in a flourishing condition. A great deal of wheat is shipped from this station—twenty million pounds last year. The town has most all the modern improvements. (Population, 2,200. Distance from San Francisco, 221 miles. Elevation, 311 feet.)

Kingsburg. This enterprising little town owes its prosperity to the fact that it is situated in the famous wheat belt. Here are to be seen big warehouses for storing wheat, large quantities of which are shipped from this station annually. The cultivation of fruit is beginning to attract attention of the people. Irrigation is the salvation of this country, and the water is secured, not only through ditches, but also by means of windmills from wells varying in depth from fifteen to fifty feet. Soon after leaving the town, we cross King's River on a trestle bridge, the approach to which is made over a long, high embankment. (Population, 450. Distance from San Francisco, 227 miles. Elevation, 300 feet.)

King's River, a large, clear body of water, rises in the Sierras to the northeast, and flows southwesterly in a broad and tortuous channel, irrigating a large scope of territory. King's River is the boundary line between Fresno and Tulare Counties.

Traver. This is a new town, showing evidence of prosperity and thrift, possesses a flouring mill, machine shops planing mills and other business enterprises of commercial importance. (Population, 600. Distance from San Francisco, 232 miles. Elevation, 291 feet.)

Goshen. The junction of the Goshen Division, which extends a distance of sixty miles to Alcalde. (Population, 75. Distance from San Francisco, 240 miles. Elevation, 286 feet.)

The Goshen Division. There are a number of small towns on this branch, as follows: Hanford, Armona, Grandeville, Lemore, Huron and Alcalde. The land through which the road passes is very fertile, and prices for it range from one hundred and fifty to three hundred dollars per acre.

Visalia. This town is the county-seat of Tulare County, and is situated seven miles to the eastward of Goshen, being connected with that station by means of a motor road. The Kaweah River flows through Visalia and aids in irrigating this most fertile region. (Population 3,000. Distance from San Francisco, 247 miles. Elevation, 290 feet.)

Resources of Tulare County. The resources of this county are most varied, the plains and the mountains meeting here; hence, the farming and fruit-raising of the one are supplemented by the mining, lumber industries and stock-raising of the other. There are about two million and a half acres of territory in the mountains, about eight hundred thousand acres among the foot-hills, eleven hundred thousand acres of valley and two hundred thousand acres in Tulare Lake and its surrounding "tule" lands. The mountains are covered with timber, and mines of gold, iron, copper and zinc are worked. The foot-hills produce almost every variety of deciduous and citrus fruits, together with grapes—both wine and raisin. Lands can be bought here at prices ranging from twenty-five to three hundred dollars an acre.

> **TULARE.**
> Commercial
> and
> Agricultural Centre.
> Population,
> 4,000.
> Distance from San Francisco, 251 miles.
> Elevation, 282 feet.

Ten miles beyond Goshen we come to Tulare, a thriving town of recent growth, with railroad round-house, shops and good station buildings, Tulare being the end of a division. This is a large shipping point, not only via the railroad, but by means of wagons to interior points.

Irrigation in the Artesian Belt. The question of irrigation in California has been one of much vexation and exceedingly difficult of solution. The supply of water has been so very limited that millions of acres of land, as fertile as any in the world if irrigated, and absolutely worthless without water, have lain fallow for years. Fortunately for California, it has been discovered that this lack of water can be supplied in many ins ances through the agency of artesian wells. In certain sections of the country these resources have been developed, and the result has been the establishment of what are popularly known as "artesian belts." One of these zones extends from Calienta to Stockton, the greatest development being in Merced, Fresno, Tulare and Kern Counties, where over seven hundred flowing wells have been established. These wells are from 250 to 700 feet in depth, and an average well will irrigate about 150 acres of land. The capacity of each well can be largely increased by means of storage reservoirs. After leaving Tulare the derricks of artesian well-borers can be seen on each side of the railroad in great numbers.

Tipton is a small station of no very great importance, except from the fact that it is the shipping point for sheep, which are raised in great numbers in the surrounding country. Seven miles to the west lies Tulare Lake, which is quite a large body of water, being thirty miles long by twenty-five miles wide, and abounding in fish and water fowl. Tipton is surrounded by a good agricultural country, and enjoys its full measure of prosperity. (Population 300. Distance from San Francisco, 262 miles. Elevation, 267 feet.)

Beyond Tipton are to be seen great numbers of windmills, used particular for the work of irrigation. Immense groves of eucalyptus, or blue gum trees can be seen from the train. Pixley, Alila, Delano, Poso and Lerdo are small stations of minor importance. We cross the Kern River between Lerdo and Bakersfield.

Bakersfield is the county-seat of Kern County, situated at the junction of the two forks of Kern River. The town has the usual complement of public and private buildings. It is surrounded by an exceedingly fertile country. Fourteen miles southwest is Kern Lake, seven miles long by four wide, while six miles farther is Buena Vista Lake, a somewhat larger body of water. Irrigation has been brought to great perfection in this county, there being seven hundred miles of irrigating canals within its limits, the largest having a width of one hundred feet and a length of forty miles. Twenty-five miles southwest of Bakersfield are the Buena Vista Oil Works. This oil region, eight miles long by three wide, only needs development to become an exceedingly valuable property. Bakersfield has, as may be seen by the above, a most productive country surrounding it. (Population, 2,500. Distance from San Francisco, 314 miles. Elevation, 415 feet.)

Caliente. This station is at the entrance to the famous Tehachapi Pass, and is located in the embrasure of a deep and narrow cañon, up which the train takes its difficult way. This is a shipping point for freight from interior points. delivered to the road by wagons. It is also quite a stage station, stages leaving

THE LOOP.

Caliente for Basin, Havilah, Hot Springs, Weldon and Kernville. (Population, 50. Distance from San Francisco, 336 miles. Elevation, 1,290 feet.)

> **The Famous Loop,**
> **TEHACHAPI PASS.**
> Distance from San Francisco, 352 miles.
> Length of Loop, 3,795 feet.
> Altitude of Tunnel, 2,956 feet.
> Altitude of Crossing, 3,034 feet.
> Altitude Gained, 78 feet.

The twenty-four miles of journey up and down the Sierra Nevadas, at the point where the railroad makes the passage of this range dividing the broad valley of the San Joaquin and the desert of Mojave, is a most remarkable experience, and brings before our eyes the wonderful triumph of railway engineering skill. It is alleged that three civil engineers of great reputation first undertook to survey a passage through these peaks and crags, and, after repeated attempts, declared the route impassible. A boy of twenty took up the work where his elders had forsaken it, and this miraculous railway path over and through the mountains is the result. Concerning this famous pass, Mr. E. McD. Johnstone writes graphically as follows: "As the Sierra Nevada and Coast Ranges in the north culminate in the great peak of Shasta (41° 24'), so in the neighborhood of Tehachapi Pass (35°) these two great chains blend their distinguishing features of fern slope and icy crag, and are lost in an inextricable mass of jumbled up peaks of every conceivable form and variety. Although nature has reared no such colossal masterpiece as Shasta in the welding of her great rock bands in the South, she has managed to throw up her earth-works in a manner so impregnable as to seemingly defy the art of man to penetrate. The physical features of this Tehachapi country (the lowest pass being 4,000 feet altitude) seemed to, and did for a time, baffle the shrewdest engineers, but, finally, the track, by doubling back upon, and crossing itself, by climbing, squirming and curving, resulted in a success and gave us one of the most famous and dextrous pieces of railroad engineering in the world."

Tehachapi Summit. The station at the summit of the pass is at an elevation of 3,964 feet, and is the highest point on this extension of the line. Sheep feed on the grass, which is abundant in the valleys and gulches which surround the station.

Descending to the Desert. For several miles the train rolls along on a level plateau on the summit of this range before the descent to the Mojave Desert is made. A small salt lake is passed, where abundance of the chloride of sodium, that important article of commerce, can be shoveled up from the bed of the lake, it being entirely exposed during the summer by the evaporation of its waters.

Cameron is a small station passed about midway between the summit and Mojave, at the base of the range

Mojave is on the edge of the desert of the same name, and the water used is brought in pipes from Cameron, a distance of ten miles. Here begins the Los Angeles Division of the railroad, and here also ends the Tulare Division. This place is the junction of the Atlantic and Pacific Railroad with the Southern Pacific. (Population 150. Distance from San Francisco, 382 miles. Elevation, 2751 feet.)

The Mojave Desert. A desert isn't as a general rule much of an object of interest to travelers, especially to those who have made the transcontinental journey and experienced the monotony of the deserts of Utah and Nevada. However we must say this, that we found many things to interest us while traversing the famed sand wastes of Mojave. In the first place there were the giant Cacti or

Yucca Palm, a sight novel to our eyes, and peculiar in and of itself. This cactus grows to the size of a tree, reaching an average height of twenty five feet, and attaining very often that of fifty feet. Its diameter is often that of two feet, and sometimes even greater; with its spreading club-like branches, its trailing bark and peculiar form, the Yucca Palm is indeed an interesting feature in the landscape. Another attraction is the peculiar form of the buttes, which rise from the desert sands on every side. Varying in height from two to five hundred feet, grooved and channeled by the elements, they give variety and interest to the landscape. One must not neglect to mention the mirage as a third element of variety. We do not remember ever to have seen more complete or deceptive mirage effects than those of the Mojave Desert.

Rosamond, Lancaster, Acton are desert stations of small interest. The Solidad Mountains tower to our right as Rosamond is passed, and we later on make our way through this range by means of what is known as the Solidad Pass, reaching an altitude of 3,211 feet.

Newhall. This station is not very large, but boasts a large hotel, capable of entertaining one hundred and fifty guests From here may be plainly seen the San Fernando Mountains, exceedingly perpendicular, and rising to an altitude of three thousand feet. These mountains could not be passed until a tunnel six thousand nine hundred and sixty-seven feet long had been made.

In this vicinity are oil refineries producing about five thousand barrels of oil per day. The oil fields are but a short distance from Newhall.

San Fernando Tunnel. From Newhall we ascend the grade through cuts until the tunnel is reached. The grade is one hundred and sixteen feet to the mile, and as we approach from the north in the tunnel, it is thirty-seven feet per mile, the grade on the south from the exit is one hundred and six feet, while the elevation of the tunnel is one thousand four hundred and sixty nine feet.

San Fernando. The valley of San Fernando bursts on our vision as we emerge from the tunnel, a land of orange groves and olive trees, the very opposite in character from the arid waste we have just left behind us. The town of San Fernando is quite a place, and growing daily in population.

Through cultivated fields, past suburban residences we roll, pausing for a moment at Burbank, only eleven miles from Los Angeles. Beyond this place we journey through villages *de facto, de jure* or *in futuro*. There are plenty of lot stakes, and the suburbs of Los Angeles will certainly be wide spread, if they ever cover the ground now laid out.

LOS ANGELES.
The Metropolis of Southern California.
A City of Tropical Magnificence.

The valley of the San Joaquin has been passed, the heights of Tehachapi have been scaled, the desert of Mojave has been crossed and we are here at last! From our cheery heights as we approach the town we gaze on a scene of entrancing beauty. Mountain-girdled, garden dotted city, lying on the slope of the Sierras, and watered by streams from the heights above, one hardly knows whether to call it a city of gardens and groves, or an immense grove and garden sprinkled with palaces and delightful homes. Health and prosperity seem to have made themselves the presiding Deities of the place. We gratefully decide that we have arrived at a point where it were well to let the train, like the busy world it typifies, pass on and away, while we rest in this paradise—a home indeed fit for the angels—and while we bask in its sunshine, gaze at its mountain peaks, catch glimpses of the ocean, breathe the fragrance of its roses and geraniums or listen to its mocking birds and nightingales, we unite many a time and oft in thanks to the kindly fate which led our steps to Southern California and the City of the Angels. There is no city whose growth can be compared to Los Angeles—in fact, no city west of the Rocky Mountains can boast of such rapid improvements. Thousands have come to Southern California simply to pay a visit, but soon become charmed with its wonderful climate and beautiful surroundings, so much so that they conclude to remain permanently in this land of sunshine and flowers. A great deal has been written of this section, but the half has never been told. With the greatest climate in the universe, the richest and most inexhaustible soil, the vast amount of valuable land in and around Los Angeles, it is no wonder that her present condition is so prosperous. The beautiful avenues extending away to the foothills on the east and to the ocean on the south, the orange groves within her corporate limits, the magnificent public and private buildings all tend to make the Angel City a place of wonder. Main street, the principal street in town, is the dividing line for east and west, First street the division for north and south. The wholesale houses are scattered along Los Angeles, Commercial, Aliso and Requena streets, while the large retail establishments are to be found on Spring street, which is to Los Angeles what State street and Wabash avenue are to Chicago. The entire city south of First street is paved with concrete pavement, north of First being laid with Belgian blocks. There are many beautiful parks within the city limits, and the ocean can be reached in a trifle over an hour's drive.

It may be stated that the much abused word "climate" has doubtless been a powerful factor in producing grand results. Furthermore, the fact that hundreds of those who were deemed hopeless invalids on their arrival here are to day enterprising, energetic and successful capitalists, merchants, manufacturers, farmers and orchardists, attesting the effects of this sun-kissed land and health-renewing climate on the human system; and so long as there are any sufferers from the blizzards, cyclones and other life destroying elements east of the Rocky Mountains, just so long will Southern California, and Los Angeles in particular, continue to receive thousands annually of the best citizens of the republic, until it becomes the most densely populated portion of the United States.

Los Angeles is reached by the Southern Pacific R. R. in twenty-two hours from San Francisco — distance, 482 miles — or by steamer. It is a most beautiful city, of 60,000 people, is growing rapidly, and is a commercial point of much importance, as well as the center of an agricultural paradise, it being the principal city between San Francisco and Kansas City on the new transcontinental line formed by the connection at Deming or El Paso. It is also the largest city between San Francisco and San Antonio, Texas, by the great "Sunset Route," now open to the Gulf of Mexico. The city has many elegant buildings, wide, clean streets, with horse, cable, and electric railways. A day's ride over the lovely country surrounding Los Angeles, through miles of long, straight avenues of orange trees and thousands of acres of grapes, seeing every kind of semi-tropic fruit growing side by side with the more hardy fruits, both being in the greatest profusion and of the finest quality, will convince the traveler from almost any part of the earth that here is surely the paradise of America, if not of the world.

No city in the United States has improved so rapidly within the past two years as Los Angeles. Since 1887 opened, nearly every one of the principal business streets have been paved with Belgian blocks, and the main residence thoroughfares with concrete, thus making a drive equal to any avenue in the Union There are no improvements which have been of more benefit to Los Angeles than that of pave-

IN THE SEMI-TROPIC ZONE, LOS ANGELES.

ment. The immense amount of daily traffic necessitated this movement, and before 1890 there will scarcely be a block within the corporate limits which will not be in proper condition. Curbing has also received its share of attention, while the cement sidewalk is becoming universal. The city has an almost perfect sewerage system, which requires an outlay of nearly $750,000. Since January 1, 1887, the Sixth Street Park, bounded by Fifth Sixth, Olive and Hill streets, has been thrown open to the public, and is in keeping with the many fine residences that surround it. The Second Street Park, situated near the terminus of the cable line of railroad, is a very inviting place, and receives its share of Eastern visitors when viewing the many improvements around Los Angeles.

Los Angeles is essentially a land of schools. The public, high and normal schools are supported by State taxation, and their doors are open to all. Besides, there are numerous universities, colleges and academies. The majority of children, after obtaining an education in the public schools, by force of circumstances are compelled to take up the battle of life for themselves; but to those who thirst for deeper draughts at the fountains of knowledge, the higher schools await them.

The University of Southern California is under the auspices of the Methodist Episcopal Church, and was established by Rev. O. S. Frambes in 1876. In 1880 it was incorporated under the State laws, and was the recipient of a large tract of land in the southwestern corner.

Los Angeles to Santa Barbara. There are two routes by which Santa Barbara may be reached from Los Angeles. One by water, via San Pedro, and the other by rail, via Saugus. A pleasant way for one with time at his disposal is by water. In order to make this trip the tourist takes the train of the San Pedro Division of the Southern Pacific Company at Los Angeles, and is soon rolling rapidly along to the southward through orange groves and vineyards, which abound along the entire course, but are especially noteworthy in the suburbs of Los Angeles.

Florence. This pretty town, embowered in an abundance of shrubs and fruit trees, is surrounded by well cultivated and fertile fields. Here the line branches, the San Diego Division extending to the left. (Population, 200. Distance from Los Angeles, 5 miles. Elevation, 151 feet.)

Compton. This is the largest town on the division between Los Angeles and San Pedro. It is in the heart of an extremely well cultivated and productive fruit belt. Grapes, citrus fruits and berries grow in great abundance. The yield is extraordinary and is especially true as to small fruits, such as blackberries, strawberries, raspberries, etc. (Population, 800 Distance from Los Angeles, 10 miles. Elevation, 76 feet.)

Ten miles beyond Compton evidences of our near approach to the grand old ocean begin to appear. Salt marshes begin to make their appearance and the fertile soil gives place to stretches of shifting sands.

Wilson's College. This is a Protestant institution of learning, eighteen miles distant from Los Angeles, situated on the site of the old Headquarters of the United States Military Department for Southern California and Arizona, which was abandoned about twenty years ago and sold to private parties. About a mile beyond the college, the junction for Long Beach is passed and San Pedro the railroad terminus is soon reached.

San Pedro. This is one of the largest and best harbors between San Francisco and San Diego. It has over a mile of docks, with between eighteen and twenty feet of water at low tide. Ships receive and unload freight to and from the railroad cars direct, though from some ships of great tonnage the freight is taken by means of lighters The government has improved the harbor to a great extent and the results have been fully commensurate with the expense incurred. The commerce of San Pedro is quite extensive, sometimes as many as twenty ships can be seen riding at anchor, or tied up to the wharf busily engaged in loading or unloading freight. Great quantities of lumber are shipped to San Pedro from points on the coast as far as two hundred miles north of San Francisco and all nations are represented during the year by ships in this harbor hailing from every part of the world. Coal comes here from the upper coast and from England and in the case of English vessels a cargo of grain is taken back. The history of San Pedro dates back to the earliest settlement of California, but as a port of any importance its growth began less than ten years ago. Before that time it was merely an open roadstead and lighters carried all freight to and from Willmington.

Point Fermin. This point is marked by a lighthouse of the first class and is one of the most conspicuous headlands on the western coast. It lies to the west of San Pedro and is reached by stage road around the beach, a distance of six miles, or by boat directly across the cove.

Santa Catalina Island. This mountainous island looms up to the southwest at a distance of twenty miles to sea. The island has become a favorite resort for excursionists and sportsmen. The trip to Catalina Island and return from San Pedro can be made for $2.00 and is well worth the visit.

San Clemeth Island is still further to seaward, faintly outlined against the sky at a distance of fifty miles. On this island great flocks of sheep and goats are allowed to range at will.

San Pedro to Santa Barbara. The tourist takes one of the steamers which ply regularly between San Francisco and San Diego and after a most enjoyable and interesting trip, finds himself at Santa Barbara, "*The Peerless.*"

SANTA BARBARA,
"THE PEERLESS."
An Ideal Home under
Sub-Tropical Skies.

Since its founding in 1786 the city of Santa Barbara has not enjoyed a more prosperous year than that just past. In the short space of two years, from January, 1887, to January, 1889, a number of events crowded together have practically changed the entire aspect of the city. At the beginning of the period, Santa Barbara was a quiet country town of great possibilities, it is true, but of very limited actual importance. It was shut off from the rest of the world, and quite contented that it should be. No changes of great moment had taken place in the town for years, and there was but little prospect that any were very soon to be realized. The marvelous advance which was being achieved by the rest of the southern part of the State showed but little signs of appearing here. People came and visited and went away, but the idea of Santa Barbara as a good place for business undertakings or for investment rarely entered their heads. Now after a brief space of time, during which the natural forces of American enterprise and the genuine merit of its situation have been allowed full play, the city of Santa Barbara finds itself a different being. Many of the old conditions still remain, some of the changes are far from complete; but those who see beneath the surface appreciate that there has suddenly come into being a new Santa Barbara with a gratifying present and a splendid future. The great event in its history, so important that it overshadows all the others, was the arrival of the railroad. Since the Southern Pacific first pushed its way into this part of the State, it has been merely a question of time when it would enter this quiet valley. On the 19th of August, 1887, the event took place, and the first train of passenger cars entered the city, bringing hosts of visitors from all parts of the State. A jubilee celebration hailed the advent of the power that was to bring progress and improvement to Santa Barbara. The company immediately set about building a suitable passenger station, and before the year was ended the railway connection had become in every sense an accomplished fact. After making its way into the city, the road proceeded along the coast, passed Goleta, and by the end of the year trains were running to Elwood, twelve miles farther north. The locality where the road enters the city, was formerly a sandy desert and is now alive with shops, freight buildings and cottages. Exact statements with regard to passenger and freight traffic are not obtainable ; but the confidence which this corporation has in its Santa Barbara connection is evinced by its purchase of more than three-quarters of a million dollars worth of property in and around this city. That the advent of the Southern Pacific Company to Santa Barbara has given a new impetus to trade and rapid advancement, none will deny.

Santa Barbara county extends along the coast of California seventy miles, and is thirty five miles in width, and has a million and a half of acres of land. It contains a population of about 30,000 and has made a gain of seventy-three per cent. during the past six years. The city is beautifully laid out with newly macadamized streets. It has electric lights, lines of street cars, telephone facilities, and everything metropolitan. It is justly termed "*The Newport of the Pacific*," with a climate unequalled for the prolongation of life, beauty and health.

> "Where the coast line trending eastward,
> Bending eastward, inward, southward
> Forms a bay of wondrous beauty,
> In a quiet, peaceful valley
> Lies a peaceful, quiet hamlet—
> Santa Barbara the peerless:
> Peerless in her genial climate,
> In her skies so clear and cloudless ;
> Peerless in her sheen of sunshine,
> Peerless, having sea and mountains,
> Shadowy cañons, mystic islands,
> Hill and valley, grove and meadow."

Malarious diseases are unknown. In fact, there are no endemic or epidemic diseases whatever. The relative humidity of the air averages seventy degrees. The average rainfall for the past few years was seventeen inches. Roses—and such roses!—bloom in the open gardens, without shelter, the whole year round, without irrigation. The city of Santa Barbara and its suburbs contains about 10,000 people. In the course of the winter it is visited by thousands of tourists from the East, which causes it to present a more metropolitan aspect than many cities five times its size. During the summer months come the visitors from San Francisco and the northern part of the State, so that at no time in the year is Santa Barbara lonesome. Horseback riding, surf-bathing, driving among the cañons and getting the views from the foothills, or merely dreaming away the hours in the calm enjoyment of the delicate atmosphere, the visitors experience no difficulty in passing the time. Santa Barbara has a future as interesting as the past has been. It is probably not destined to achieve great commercial importance. It does not expect to rival San Francisco, nor compete with Los Angeles for the first place in the southern part of the State. Through its harbor, which is one of the most perfect on the coast, it will receive a steadily increasing quantity and variety of imports ; and as the port of a rich and productive region, it must transact a considerable amount of business. But the true future of Santa Barbara lies in the manifold advantages which it possesses over other places on this favored coast, as a place for homes and villas. Not only in the town itself but in the val-

OLD MISSION CHURCH AT SANTA BARBARA.

leys and among the foothills are many perfect sites, where, surrounded by a few acres, which a little care will transform into a garden, the happy proprietor spends his days in peace and calm contentment.

The Old Mission. A visit to Santa Barbara is not complete without a meditative stroll through the old mission, the history of which is pleasantly given by Mr. E. McD. Johnstone in his delightful book "By Semi-Tropic Seas" as follows: "This of Santa Barbara is the best preserved of all the o'd missions, and has had, perhaps, the most notable history of any. Its presidio, or military garrison, was founded by Father Junipero Serra, on April 29, 1782, but it was not until the 4th day of December, 1786, on the celebration of the feast of Santa Barbara, virgin and martyr, that the cross was raised and the mission founded. A few days after, the Rev. Father Lasuen celebrated mass and preached from a hut or booth made for the occasion. The territory under control of this mission included all the arable lands from the 'Rincon' west to Point Conception, and from the mountains, on the north, to the sea. The greatest prosperity of the mission was reached about 1812. The unjust demands continually made upon it by the Spanish government, and later by the Mexican, greatly weakened its resources, and, finally, by the secularization act and the withdrawal of Mexican protection, the destruction of this property, as far as its primal object was concerned, was complete. The immediate property of the mission was leased in 1845. In 1852 it was organized into an independent Franciscan convent or college. In 1885 this college, the titular of which was the Blessed Virgin of the Seven Dolors of Santa Barbara, was annexed to the Province of the Sacred Heart of Jesus of the U. S. A."

By Rail to Los Angeles. As a pleasant variety the tourist can make the return trip from Santa Barbara to Los Angeles by rail, taking the Ventura Division of the Southern Pacific Company, which forms a junction at Saugus with the main line from San Francisco. Leaving Santa Barbara with a sigh of regret, for here, if anywhere, a man could live peaceful days, we pass through groves of verdure and are soon skirting the ocean with towering cliffs to the landward, which in places have been blasted away leaving space for the railroad to pass. The scenery is charming and the mind is pleasantly engaged until the train pauses at a handsome station building, which is the depot for

Ortega. This is a town of great expectations, with plenty of lot stakes in sight, but few buildings. The situation is a charming one and there seems to be every reason to believe that ere long a town worthy of the beautiful, natural surroundings will be established. (Distance from San Francisco, 521 miles. Elevation, 77 feet.)

Carpinteria. This is a picturesque little town surrounded and encroached upon by orchards and vineyards. Many of the residents are of Spanish origin, as Carpinteria dates back to the early settlement of Southern California and was one of the original m'ssion towns. (Population, 300. Distance from San Francisco, 517 miles. Elevation, 8 feet.)

San Buenaventura. The city of San Buenaventura, since the advent of the Southern Pacific Company, has made progressive strides. It is a beautiful, old, ex-Spanish town, with 3,000 population, and is the county seat of Ventura county. It is beautifully located upon the seashore, just at the point where the Ventura river breaks through the sand into the sea. There are not many costly or elegant residences, but rather more than the usual proportion of neat, cosy homes. The same thing is true of the business blocks, and the town has its full complement of stores and shops of all kinds. The streets have solid concrete walks and tile sewerage throughout the town. It is well lighted with gas, and has very efficient water-works. The hotels are good and well patronized. There are the usual number of churches, schools and fraternal organizations. The town is lively and a large volume of business has always been transacted. The climate is delightful and life here must be a continual delight.

> " And oh ! the balmy air 'tis bliss to breathe,
> As through the mountain gap steals the fresh breeze,
> Tempering the fervid summer's noonday heats
> With the gentle breath of mild Pacific seas."

San Buenaventura has until within a comparatively recent date, been, in a measure isolated from the great centers of trade, not only of California but of the world ; but now that the Southern Pacific Company has been built through its confines, it has entered upon an era of unexampled prosperity. The road enters the southeastern part of the county by the way of Newhall, and extends to the coast at San Buenaventura, tapping a tract of country that for fertility cannot be surpassed. Los Angeles is but a few hours' ride distant ; and the varied products of this section find a ready market. (Population, 3,000. Distance from San Francisco, 500 miles. Elevation, 45 feet.)

Fertile Valleys. The great Santa Clara of the South is celebrated for the fertility of its soil, mildness of climate, and healthfulness of its people. It was this valley and its tributaries that enabled Ventura county at the Mechanics' Fair at San Francisco, in 1885, to carry off the first premium for the most extensive and varied exhibit of farm products. The valley extends nearly east and west across the county, and is traversed by the Santa Clara river, fed by numerous tributaries, as the Castic, Piru, Sespe and Santa Paula. At the upper or east end is the San Francisco ranch, which includes the wheat-growing ranch of the Newhall Brothers, and the Camulos with its orange and olive orchards, wine-cellars and old vineyards, made famous by Mrs. Jackson, who wrote a part of her celebrated book, " Ramona" at this place. San Francisco ranch contains about 12,000 acres ; the Sespe ranch 8,000, well adapted for citrus and deciduous fruits ; the Saticoy Rancho, 17,000 acres ; Bardsdale, a 2.000-acre colony ; thence southerly, the Colonia rancho of 45,000 acres, level as a floor. Intervening is the Rancho Santa Clara del Norte, of 13,000 acres. Next there is the San Miguel rancho, of 5,000 acres—an immense corn, bean and grain field. When we have passed through this rancho, en route for Los Angeles, we are in the Ex-Mission, the grant upon which is located the beautiful and thriving palm city, San Buenaventura. This rancho includes about 48,000 acres, mostly hill lands, lying north of Santa Paula and Saticoy. Among its hills are some beautiful tracts, well wooded and watered. In this great valley, the Santa Clara of the South, a large population can be sustained. Its wonderful resources, climate and scenery attract the attention of home-seekers.

The Ojai valley is a great amphitheater, whose walls are mountains rising like citadels in all directions. Overlooking the whole is Mt. Topo-topa, rising to a height of from five to six thousand feet, and coming out in springtime from the snows of untold winters as fresh and beautiful as ever. The drive to lower Ojai is exceedingly inviting, being an easy grade along a clear, beautiful stream alive with trout. In many places the road is arched with sycamore, oak and other trees, festooned with hanging mosses and vines, and made vocal by the songs of birds. In the valleys the air is soft and balmy as that of the island of Atlantis of fabled story. They are the resort of invalids and pleasure-seekers, who receive the best attention and care at a very moderate price.

Montalvo is another prospective city, with great beauty of situation, really a suburb of San Buenaventura, but no great business interests at present. (Distance from San Francisco, 495 miles. Elevation, 89 feet.)

Saticoy. This pretty little town is situated in the midst of a fruit, grain and vegetable growing region. Population, 95. Distance from San Francisco, 491 miles. Elevation, 146 feet.)

Santa Paula is one of the leading towns in Ventura county and is fifty miles north of Los Angeles. It has a growing population and is located in the center of the beautiful Southern Santa Clara valley, which has the most productive soil in the world, producing anything that mother earth can bring forth. Grain, corn, beans and tropical fruits are raised in abundance, and are unsurpassed in quantity and quality. There is one orange orchard near the town consisting of one hundred acres, which is the finest we ever saw.

For climate and health, Santa Paula and its surroundings are unexcelled. The water supply is abundant from cooling springs in the near mountains. The fine gardens of vegetation and flowers ripen and bloom the whole year round. In fact, it is a land overflowing with milk, honey—and oil.

Santa Paula is the headquarters of the oil regions of California. The most extensive Petroleum Oil operations are on the Rancho Ex-Mission, situated along the south side of Sulphur Mountain, beginning about four miles northwest of the town and extending in a westerly direction eight miles, these wells are owned and operated by a company which is incorporated with a capital stock of $1,000,000. This company has been most successful in its development, having a daily production of about 1,000 barrels from the many wells and tunnels. The region is a network of pipe lines conveying the oil to Santa Paula, Ventura and Hueneme. The largest well produces about 300 barrels daily. The next most extensive oil developments in this region are located at the Sespe, owned and operated by the Sespe Oil Company, with its office at Santa Paula. The company has a capital stock of $250,000. The production of the region is about 275 barrels daily, which is piped to Santa Paula. These two companies keep a large force of men constantly engaged in drilling new wells, and thus the production is being constantly augmented. No industry in the Golden State promises better results than its oil developments, and nothing is more beneficial to Ventura county, and to Santa Paula in particular. With an abundance of cheap petroleum for fuel, no section offers better advantages for manufacturing purposes. (Population, 900. Distance from San Francisco, 483 miles. Elevation 286 feet.)

Camulos. This picturesque hamlet has been made known to the world of book readers as the home of "Ramona." The scenery surrounding it is of the most attractive character. The San Fernando mountains are on the south, the foot-hills of the Sierra de San Rafael on the north, the Santa Clara river flows through the sylvan valley that lies between. On its margins are clumps of willows and groves of wide-spreading sycamores, and near where its clear waters run by the old homestead, may be seen the "artichoke patch," and the "flat stone washboards, on which was done all the family washing." The house, as described by Mrs. Jackson, was "one of the best specimens of the representative house of the half barbaric, half elegant, wholly generous and free-handed life led there by Mexican men and women of degree in the early part of this century." The foot hill pasture lands, the sheep corrals, the vineyards, olive groves and orchards, the old Chapel, etc , etc., are all to be seen quite as really as they are described in this interesting book. Mrs. Jackson's descriptions of Southern California scenery are exceedingly fine, and it is not a matter of wonder that she chose this beautiful spot as the home of her charming Ramona. Camulos presents opportunities for the establishment of ideal homes in the heart of ideal scenery. (Population, 150. Distance from San Francisco, 463 miles. Elevation, 286 feet.)

Saugus. Junction of the Ventura Division with the main line. Our journey from here to Los Angeles has already been described.

From Los Angeles to San Diego. The trip from Los Angeles to San Diego abounds in interest and if one obeyed one's inclinations and made a stop at all the attractive stations which intervene between the inland city and the city on the ocean side it would take an entire vacation to accomplish the one hundred and eighty-nine miles of the journey. Leaving Los Angeles on the California Central Railway at a comfortable hour in the morning, we are soon speeding through the suburbs of the City of Angels. It is difficult for us to tell just when we have passed beyond the confines of the city, because the country is so fully occupied by handsome villa residences and the suburban stations are of such frequent occurrence that one is puzzled to determine where the town ends and the country begins. Downey Avenue, Morgan, Highland Park, Gravanzo, Lincoln Park, South Pasadena, Raymond, Pasadena, Olivewood, Fair Oaks and Lamanda Park are all busy stations disposed within a distance of thirteen miles from Los Angeles. It is therefore not to be wondered at that the traveler is confused and at a loss to know just when he is "out of town." Beyond Lamanda Park the stretches of open country between stations begin to widen and one can look out of the window at least twice before another town appears in view.

Raymond. As this station is approached one sees on the right an aspiring hill adorned with handsome lawns, ornamental shrubbery, trailing vines and umbrageous trees. The summit of this hill is crowned by a massive and stately edifice that at once attracts attention and excites curiosity. On inquiry we learn that this is the Hotel Raymond and that here are entertained the hundreds of guests brought hither by the well known excursion managers, Messrs. Raymond and Whitcomb. This, however, forms but a small part of the patronage of the Hotel Raymond, for from its excellent management, beautiful situation and healthful location the hotel has become exceedingly popular. Of course there is a town-site here and, what is not always the case in this country of town-sites, there is a town as well, with the prospects of a city.

South Pasadena is a flourishing suburb of Pasadena and will soon be so merged into the parent town that they will be practically one and the same city.

PASADENA.
An Orchard City.
Beautiful for Situation.

A Delightful
Health and Pleasure
Resort.

One of the loveliest towns in the world lies before us as we enter Pasadena. From a sheep range in 1873 to the paradise of fruits and flowers and verdure which greets our eyes to day is a magic transformation. Yet such, in a word, is the history of Pasadena. The semi-tropical luxuriance of floral and arboreal growth which delights us here has sprung into existence within the marvelously short space of a decade and a half, and, nestling here among the orange groves and fruiting vineyards, is a city whose beauty of architecture is a glowing testimonial to the good taste, wealth and liberality of its residents. I know of no pleasanter or more interesting drives than those which may be taken along the broad tree-lined avenues of Pasadena. Within spacious enclosures on each hand may be seen elegant villa residences or splendid mansions surrounded by ornamental grounds of the greatest beauty. Palm trees, magnolias, century plants, fig trees, ancient live oaks, survivals of the days when this was only grazing ground for flocks and herds, pepper trees, blue gums and an infinite variety of ornamental shrubbery, make these drives entirely novel, interesting and charming. The city obtains an abundant supply of water from the Arroyo Seco Cañon and the results of irrigation confront one in the wonderful groves of citrus and deciduous trees. Pasadena has a round dozen of churches, representing an expenditure of nearly half a million dollars. It has business blocks of metropolitan proportions, spacious and elegant theatres,

NEAR SAN GABRIEL AND PASADENA.

four banks, a score of hotels, large manufacturing establishments, canning factories, horse car lines, telephone system, electric lights, — in short, all of the modern conveniences. As a place of residence we know of no more charming city than Pasadena, whose ten thousand inhabitants have every reason to congratulate themselves that their lines have fallen in such pleasant places. The wonderful climate of Pasadena is one of its chief attractions. Tourists who arrive in November or

October are constantly on the watch for winter. Finally a rain storm comes, drenching the earth, and a few weeks later the ground the length and breadth of the land is carpeted with flowers, form succeeding form, until color and variety, tint and hue, seem to have run riot; by this token you may know that the winter has come. The tops of the Sierras are clothed with snow, so near that you can see the snow blown high in air by the mountain's blizzard, so near that in two hours' ride you can go snow balling or tobogganing, yet here at Pasadena the ground is white with the blossoms of the orange, there is a carnival of flowers in every dooryard, and to the student who arranges his plants according to their altitudinal horizons, it is a puzzle. Here, in the same latitude as Wilmington, N. C., we find the banana, fig, pomegranate, guava, alligator pear, cocoanut, the fan palm, sago palm, cactus, the yucca, century plant, cork tree, the rubber tree, the olive, orange, lime, lemon, and a host of other tropical forms, yet it can not be a tropical climate, as side by side with these is seen every pine known from Norfolk Island to the shores of the Arctic Sea, firs, spruces; and as for fruits, we see the apple, pear, peach, apricot, plum, nectarine, all the small fruits, everything found in the gardens of New York State.

The seasons are difficult to understand. The summer mean temperature at Pasadena is 66.61 degrees; that of Mentone in the Riviera, 73 degrees; of Jacksonville, Fla., 81 degrees; of New York, about 73 degrees. Thus it will be seen Pasadena can not have remarkably warm weather. The summer, with the exception of one or two days, is not unpleasantly warm, and it is always pleasant and comfortable in the shade, while every night is sufficiently cool to require a blanket. Not a case of prostration from heat, not a squall or wind storm, seldom a thunder-clap or sign of lightning, and hardly a cloud in the sky; this is the record of the summer here. Every day is a pleasant one, and such heat as is experienced in New York City in the summer is never felt.

Three hundred and forty days out of the year will permit of continuous out-of-door life in the open sunlight, and at least half of the others may be enjoyed. This is the great secret. The country is the land of the open air, winter and summer, and the conditions of altitude and nearness to large cities allowing of all the luxuries and comforts, add to its attractions.

For further descriptive matter concerning this place the reader is referred to the addenda.

Lamanda Park. We wish to do the tourist who reads this book a good turn, having his comfort and enjoyment at heart; therefore we advise him to stop at Lamanda Park and make his headquarters for a day, or a week, or a fortnight, in this delightful spot. In the first place one can find here a home-like and comfortable hotel; in the second place, this is an excellent point from which to make radiating trips through the charming San Gabriel Valley or among the foot-hills and up the peaks of the Sierra Madre Mountains. Within an hour's drive are Sierra Madre Villa, the famous Rose Vineyards, Baldwin's ranch, where, besides miles of orange avenues are to be found, at Santa Anita, the stables made famous by the fast horses owned by the "bonanza king." Orange orchards, avenues of English walnut trees, lemon groves, vineyards, veritable forests of deciduous fruit trees and a tropic luxuriance of splendid floral beauties surround this place, which, though modest in size, is as we have said, a charming resting spot and a most convenient point from which to radiate in all directions and view either the grandeur of the mountains or the more quiet but none the less attractive beauties of the valley. For further descriptive matter concerning this place the reader is referred to the addenda.

SIERRA MADRE VILLA

An Ideal Pleasure
and
Health Resort,
In the Heart of Orange
Groves, on the
Slope of the Sierra
Madre Mountains.

The fame of the Sierra Madre Villa is world-wide. On its shaded verandas congregate daily the most cultivated and intelligent people. It is not always the same company that gathers here, but it *is* always a company which it gives pleasure for one to meet. The class of guests is of the best, because the reputation of the Villa naturally attracts that class. This ideal pleasure and health resort is located on the southern slope of the Sierra Madre Mountains, fourteen hundred feet above the level of the sea. This elevation gives it complete immunity from the fogs of the sea and valley, and also gives a view of the most wide-horizoned beauty. Here we are only fourteen miles from Los Angeles, far enough away to escape the turmoil of the city, and near enough to enjoy all of its advantages. Theatre trains are run three or four nights each week, and one can go to Los Angeles by train at almost any hour in the day. The California Central Railroad passes within a mile and a half of the villa, Lamanda Park being the station. The views from the Villa overlooking the beautiful San Gabriel Valley, are a glorious panorama of rugged mountain ranges, extensive orange groves—in one of which the Villa stands—vineyards, and the distant ocean with its shadowy islands. Here is, indeed, an ideal home. with good food skillfully prepared, pure air and sparkling mountain water. With all these essentials for health, comfort and luxury, the tourist can not fail to enjoy his sojourn here. The fame of the Villa for its beautiful and healthful location, and superior accommodations, with all modern improvements for over one hundred guests, has become international. There are fine suites of sunny rooms, broad verandas, inclosed with glass to keep out chilly air if desired, a beautiful lawn, flowers, etc., and the most genial climate under the sun. Good roads and a beautiful drive from Los Angeles to the Villa. Eight trains daily leave Los Angeles from First street depot, all stopping at Lamanda Park, where the stage from the Villa meets all trains.

The San Gabriel Valley. The remarkable growth of the San Gabriel Valley of Southern California may be traced to a single imperishable feature — its climate. Towns and cities have appeared like magic; not the mushroom growth one expects and finds where a mining excitement has been the magnet, but towns which in completeness, architectural beauty, taste and culture of the people, will equal many in the East dating back fifty years or more. Ten years ago the San Gabriel Valley was, comparatively speaking, unoccupied. Several small towns, as Duarte, San Gabriel, Puente, were the chief centres, and the entire land was cut up into large holdings or ranches. To day we find towns by the dozen larger than these pioneers, three lines of transcontinental railway, and one city, Pasadena, with a summer or permanent population of fifteen thousand persons, and a winter one ranging from twenty thousand to forty thousand. The San Gabriel Valley is about ten miles wide and thirty miles long. Upon the north are the California Maritime Alps — the Sierra Madre range — rising directly from the plains in a series of parallel ridges, in peaks from four thousand to fourteen thousand feet above the sea. To the west, spurs of the main range, the Sierra Santa Monica, the San Rafael and the Verdugo Mountains form a protective boundary, while to the south the Puente Hills rise, beyond which, faintly visible, twenty-five miles away, is the Pacific. The Valley is therefore completely environed on all sides, having absolute protection from prevailing winds from the north, in this respect again resembling the Riviera of Europe. The presence of these mountains and cañons rising so abruptly from the valley gives to the locality a scenic charm difficult to describe, and for its peculiar charm the view of the Sierra Madre range at Pasadena is unequaled in this country.

Monrovia. This handsome little city has been christened by its admirers "The Gem of the Foot-hills," and, in fact, there is quite as much truth as poetry in the title. It has a most attractive site, commanding a comprehensive view of the San Gabriel Valley to the front, while the background is filled in with the massive range of the Sierra Madre mountains. The town has two lines of street railway and a motor line to Los Angeles was nearly completed at the time of this writing. It possesses an elegant and costly hotel, furnished with all the modern improvements, handsome school-houses, first class business blocks, fine private residences, and no saloons.

Duarte. This is one of the oldest of the settlements of the Valley and is surrounded by a country of great productiveness. Farming is a considerable industry and great quantities of corn and alfalfa, in addition to fruit, are raised.

Azusa is near the upper end of the San Gabriel Valley and is in the centre of the great ranch from which it takes its name. The stations now follow in quick succession until San Bernardino is reached. In fact the train never makes more than four miles advance without either stopping at a station or passing through one. To give the reader an idea of the frequent occurrence of these towns we append a list, with the distance of each from Los Angeles: Glendora, 27; San Dimas. 31; Lordsburg, 34; North Pomona, 35; Claremont, 36; North Ontario, 41; North Cocamonga, 45; Etiwanda, 47; Rialto, 57, and San Bernardino, 60. One of the most marvelous things connected with this journey of sixty miles from Los Angeles to San Bernardino, aside from the marvels of nature, is that for an average of every three miles of the journey there is a station and that at many of these stations there are considerable towns and at several of them thriving cities.

> **SAN BERNARDINO,**
>
> Manufacturing and Mercantile Centre.
>
> ---
>
> A Beautiful Residence City.

At San Bernardino we stop for dinner and change cars, taking the California Southern Railroad for San Diego. The station is a large and spacious building, admirably fitted for the purposes to which it is dedicated. The city of San Bernardino lies in a most beautiful and fertile valley. The county embraces 23,476 square miles, and contains not only some of the finest farming land, citrus and deciduous fruits of countless varieties, but also rich mines, and many mineral springs and health resorts. The scenery is magnificent and varied, the mountains abound in timber, and game is plentiful. The climate is superb and invigorating. The city of San Bernardino is situated in the centre of a valley one mile square and has a population of 10,000, and is rapidly increasing in size and wealth. Among other notable buildings are the finest brick grammar school in Southern California, a court-house which cost $40,000; the Stewart Hotel, costing $125,000; an opera house, an excellent hospital, and churches of all denominations. There is an abundance of artesian water. Three lines of railroads cross the county—the Southern Pacific, through Los Angeles and Colton, to Yuma and Arizona, and the California Southern from San Diego, through San Bernardino to Barstow, where it connects with the Atlantic & Pacific; and the Los Angeles & San Bernardino Railroad, which runs on a straight line between the two cities. There is also the Valley Railroad, from the city to Gladysta, Lugonia, Redlands and Mentone, to the west line of High View.

A n. tor road also runs continuously between this city and Colton, a distance of three miles, and the San Bernardino & Redlands Motor Road to Redlands and Lugonia, via Victoria and Old San Bernardino orange groves. Also the San Bernardino & Arrowhead Narrow Gauge to Arrowhead Hot Springs. Street cars are running to all parts of the city. Building material is abundant and cheap. Among the varied products that attain perfection here we may mention oranges, raisins, wines, fruits and flowers of all kinds, alfalfa, corn and barley, while gold, silver and borax are found in large quantities in the near mountain ranges.

A Fertile Valley. The county of San Bernardino is the largest in California, and includes within its limits the valley of the same name. It contains much land which is now lying fallow, but which will in time be irrigated and made very productive. In its southwest corner are several large valleys well irrigated and of unusual fertility. Within them are long stretches of almost level plains, from which the gently undulating mesas gradually rise until they reach the foothills. The lower level lands are sufficiently moist to grow alfalfa, corn and vegetables, without irrigation; and the soil is mainly a black sandy loam. The higher lands become more sandy, while the foothills contain the gravel washings from the mountains. These higher lands grow vines and deciduous fruits with the natural moisture; oranges and lemons alone require artificial irrigation. The higher lands are better for deciduous fruits, the mesas or table-lands for citrus fruits, the lower lands for vegetables and general farming. There are some immense vineyards in the country, and a vast quantity of excellent wine is made. After a barley crop is harvested, it is succeeded on the damp or irrigated lands by a crop of corn. Alfalfa yields well and is cut from three to seven times in the season. About two tons are taken off each acre at a cutting. The heavy black loam of the mountain sides grows exceptionally fine potatoes. Vegetables and edible roots of all kinds attain an enormous growth in the valley. Besides the semi-tropical fruits, all those of more northern latitudes can be raised. These valleys surpass any others in the southern part of the State in the matter of an abundant supply of water for irrigating purposes. The Chino Ranch and Ontario lands are in this county, as well as those of Riverside, whose oranges and raisins have gained a National reputation.

Colton. This live town is at the crossing of the California Southern and the Southern Pacific railroads, and an unusually handsome station and large hotel are to be seen here. The town is only four miles from San Bernardino, and the time is not far distant when they will be one city. The citizens of Colton are enterprising and liberal, and as a result the town is making rapid and large improvement. Canning factories are established here, and the shipments of prepared fruit and fruit in its natural state are something extraordinary. The surrounding country is of unsurpassed fertility, and a drive of half a day through the never-ending groves of orange trees and in the midst of most entrancing scenery will convince one that Colton has every requisite for becoming a large and flourishing city. It is surely a most delightful place of residence. For further descriptive matter concerning this city the reader is referred to the addenda.

East Riverside is the station for Riverside, reached by a branch line.

South Riverside, on the California Southern Railway, 15 miles southwest of Riverside, is remarkable for the beauty of its situation and the symmetry of its design. The projectors of this delightful town had original ideas and the town-site is exactly circular in form. Fruit raising is one of the leading industries, while manufacturing is receiving a great deal of attention and has already been firmly established here. For further description of this colony see addenda.

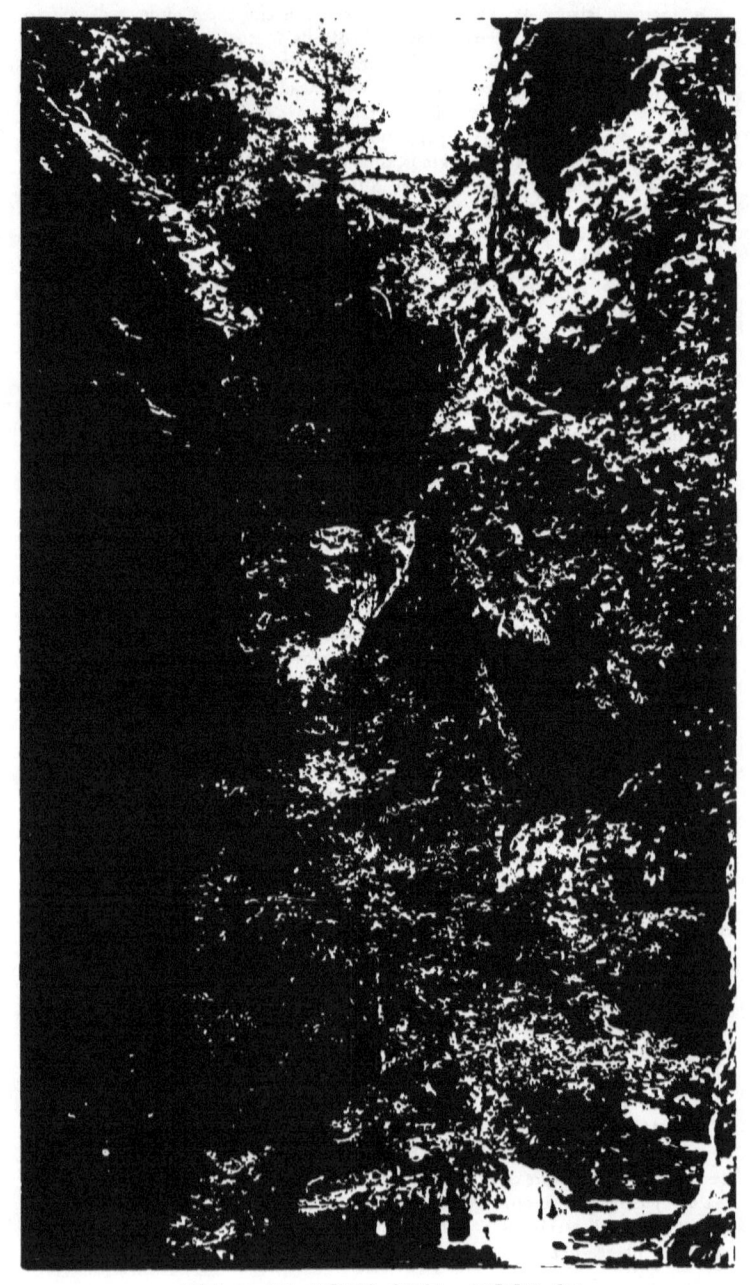

VIEW IN SAN ANTONIO CANON, ONTARIO, CAL.

RIVERSIDE.

The Orange Grove City of Southern California.

Washington has been wittily denominated "the city of magnificent distances," but here in Southern California we have found a city equally as deserving of that characterization. Riverside manages to cover twenty-five thousand acres, and this great extent of territory has upon it between three and four thousand inhabitants. But did ever anyone behold a more beautiful sight than this orchard city, reclining in the midst of orange groves, its magnificent avenues lined with ornamental trees, among which the oriental palm is most conspicuous, its artistic villa residences surrounded with grounds in which the care of the landscape gardener, can be seen, its fine business blocks of brick and stone, its handsome hotels and its surrounding vineyards making it a perfect bower of beauty.

Resuming our journey on the main line from East Riverside, we pass through Box Springs, Alessandro and Perris, which latter place is situated on the San Jacinto River, which empties in Lake Elsinore, some twelve miles farther on. The country has become more rugged, for we are now skirting the San Jacinto hills. We pass through deep cuts and around projecting spurs, and finally enter a very pretty cañon, emerging from which we pause at Elsinore on the margin of

Lake Elsinore. This is a lovely little sheet of water, cradled in the highlands, with a bold mountain range to the west. The lake is four miles long and about half a mile wide, and forms a charming feature in the landscape.

Wildomar. At the foot of Elsinore Lake is Wildomar. This town has a very picturesque situation, and considerable expense has been ncurred in planting trees, grading the streets, and bringing water in pipes from the adjacent mountains. It has schools, churches, good business houses and a population of about two hundred. For further descriptive matter concerning this place, the reader is referred to the Addenda.

Murietta. This is a regular meal station, and on that account is of interest to the traveler. It is situated on the Margurita ranch, which comprises 208,000 acres of land, especially and solely adapted for grazing. San Margurita Creek flows through the town, and the railroad follows this stream for thirty seven miles, and then, over the brow of a rolling mesa to our right, the great Pacific Ocean bursts on our view.

Ocean Side. This thriving town of a thousand inhabitants has a commanding situation on a mesa two hundred feet above the level of the ocean. From this point of view the coast line can be followed in either direction as far as the eye can reach. Here there is one of the finest hotels (The South Pacific) on the coast, and here great improvements have been inaugurated by the enterprising citizens. The accommodations for sea bathing are most complete, and Ocean Side is sure to become an exceedingly popular pleasure resort. Between Ocean Side and San Diego, a distance of forty seven miles, there are just a "baker's dozen" of stations. At some of them one can see hotels of the most imposing size and beautiful architecture, a house or two, and thousands of lot stakes, but no great showing of business or population. The stations occur in the following order: Carlsbad, Leucadia, Encinitas, Del Mar, Cardero, Sorrento Alpine Selwyn, La Jolla, Roses Siding, Morena and Old Town.

SCENE IN SAN ANTONIO CANON: ONTARIO, CAL.

SAN DIEGO.

The Naples of the New World.

The Great Bay City of Southern California.

The magnificent natural advantages of San Diego cannot fail to make this the great city of Southern California. It lies upon a slope facing San Diego Bay. This slope extends back perhaps an average mile, where it reaches an altitude of 200 feet above the level of the sea, and from which point the country extends back in a broad, rolling mesa. With such a slope, and with such an ascending altitude, opportunities are offered for the most wide-sweeping and magnificent views. At the foot of the city lies the land-locked bay, one of the most beautiful in the world, glistening like a sheet of silver in the genial rays of an unclouded sun. Between the bay and the ocean is the Coronado peninsula, on the expanded part of which is the town of Coronado, with the largest hotel in the world. Beyond Coronado is the Pacific Ocean, whose long, rolling swells break upon a level and far-extending beach, their combining crests breaking into snow-white foam as they fall with majestic regularity upon the shining sands. The distant background is formed by the mountains, with the Jamul, old San Miguel and El Cajon standing well forward, the advance guard of an army of giants. To the right is the receding mesa; to the left the table lands and mountains of Old Mexico. The landscape in garb of varying green, the bay and ocean with their ever-changing shades from shining silver to deep, dark blue, form a picture of such entrancing beauty that neither pen nor pencil can adequately depict. With such natural attractions, to which should be added the attractions of climate, it is not a matter of wonder that the population of San Diego has increased rapidly since overland transportation facilities have been provided. The city's population in November, 1885, was but the population of a healthy village, say about four thousand; a year later saw it advance to a city of between ten and twelve thousand; and by November, 1887, the population had doubled again, and reached a total of twenty-five thousand souls. The increase since has been steady, and the common but conservative estimate of the population to-day is thirty thousand. The character of the population is truly American. Because to the Eastern mind San Diego is "away in the West," the impression prevails with some that its population is of that western character to be found in romance of the light order. A greater mistake could not be imagined. San Diego is as typical an American city as any to be found in the land of Americans. If the influence of any one city may be said to prevail here, it is the influence of the City of Boston; and there is reason for it. The Sante Fé Railroad, whose western terminus is at this harbor, is an institution maintained by Boston men and Boston capital. This has naturally created in Boston a financial, and finally a social, interest in San Diego, which has resulted in the transplanting of many Boston men and women from the metropolis of New England to the new city by the sunset sea. They have found here a genial, social climate. In a city covering as much ground as does San Diego, the matter of transportation is of first importance. This has been looked after by the enterprising citizens. Horse cars, steam motors and electric motors are already in use, and a franchise for a cable system has been granted, upon which it is expected work will begin in a short time. The San Diego Street Car Company has in operation twelve and one-half miles of horse-car lines. The Coronado Railroad Company has in operation twenty-eight miles of suburban steam motor lines. The Electric Rapid Transit Company controls about six miles of road, and is rapidly extending its lines, which, under the Henry Electric System, are being operated with great success. The National City

& Otay Railway Company have twenty-nine miles of steam motor lines and three-fourths of a mile of horse car line under operation. These lines centre in the city, and afford frequent and rapid communication to all parts and to the suburban towns and valleys.

SAN DIEGO BAY.

A Thing of Beauty and a Great Commercial Factor.

The bay of San Diego is one of the most beautiful in the world; it is also a great factor in the success of the city. There are larger harbors than this, but for the uses to which harbors are devoted, there are none better anywhere than that of San Diego, and it is large enough to afford a safe refuge for the entire merchant fleet of the United States. The bay is thirteen miles long, and the total area of water is twenty two square miles. Commodore C. P. Patterson, Superintendent of the United States Coast Survey, wrote in 1878:

"I have crossed this bar at all hours, both day and night, with steamers of from 1,000 to 3,000 tons burden, during all seasons of the year, for several years, without detention. It is the only land locked harbor south of San Francisco and north of San Quintin, Lower California, and from a national point of view its importance is so great that its preservation demands National protection, and justifies National expenditure."

It may be added right here, however, that the Government has never acted upon Commodore Patterson's worthy suggestion Not one dollar of Government money has been spent in either the improvement or preservation of San Diego harbor. It stands to-day as nature made it. The depth across the bar is 23 feet at mean low water, with a rise of from 3¼ feet to 5½ feet at high water; and a regular trader at this port, the ship "Jeremiah Thompson," drawing 26½ feet of water, comes into the harbor and reaches the wharves without danger or difficulty, bringing an average cargo of 1,500,000 feet of lumber.

The history of this wonderful city reads like a romance. Previous to November, 1885, San Diego existed chiefly as a town-site, and, measured by the corporation limits, it contained an amplitude of area. It was in 1833 that the Pueblo of San Diego was organized; but it was not until eleven years later, in the latter part of the year 1844, that the people followed the usual customs of those times, and petitioned the Government of Mexico (this whole country was then under Mexican rule) for a tract of land. A few acres more or less was of no particular account to the Mexican Government at that time, and a grant of seventy five square miles was made, "to be used, controlled and disposed of by the legally authorized representatives of the city." These seventy-five square miles or, to be exact and use the figures of the surveyor who traced the lines subsequently for the Government, and who reported that the entire Pueblo consisted of 48 556.69 acres, do now, minus 1,233 8 acres reserved by the Government for military purposes, constitute the area of the corporation of San Diego. The question of title never arises here. That original grant has been confirmed, and upon it rests all instruments of sale.

The shores of the bay are dotted with suburban towns, which share the benefits of San Diego harbor. They are separated from the City of San Diego by distinct bounds, but it is only a matter of time when they will become integral parts of the parent city. These towns are known as Nat onal City, Roseville and Coronado.

National City is located four miles down the bay, reckoning the distance from the center of the business community of each city. The two cities are, however, already practically merged into one, as they are one in interest and in sentiment. National City has a population of 3,000. It is the terminus of the Santa Fé system on the Pacific coast, and of the National City & Otay Railway Company. A capacious wharf furnishes facilities for deep-sea vessels to unload, and here, too, ship and rail are brought together. An olive oil mill having been established, National City is the olive market for Los Angeles, San Bernardino and San Diego counties. It is furnished with water from the recently completed Sweetwater reservoir, which has a capacity of six billion gallons, and insures a supply sufficient for a city of twenty-five thousand inhabitants.

Coronado. On Coronado Beach, just across the bay from San Diego, is a city which has already become famous throughout the country. In two years' time this wild waste of land has been transformed into a city with a population of two thousand. It has one hotel which cost one million dollars, and others which cost large sums; it has elegant and substantial residences; it has an iron foundry in

YOUNG AMERICA'S FRIEND.

operation, and half a dozen factories of various kinds; it has ship-ways with a capacity for dry-docking the largest coast steamers on an hour's notice; it has complete water, gas and sewer systems, and, as a whole, has been converted into a veritable garden, the streets being uniformly lined with tropical trees, shrubs and flowers. The surf-bathing of Coronado Beach is the best on the entire coast, and probably the finest in the world. The beach slopes gently, and the sand is hard and free from stones and ragged shells, and there is no undertow. The temperature of the air and of the sea is about equal both in winter and summer; consequently there are extraordinary inducements for surf bathing all the year round.

Roseville and New Roseville are located not far from the entrance to the harbor. A fine wharf has been built there, regular ferries established, and the works of the San Diego Nail Factory are now being erected. They will have

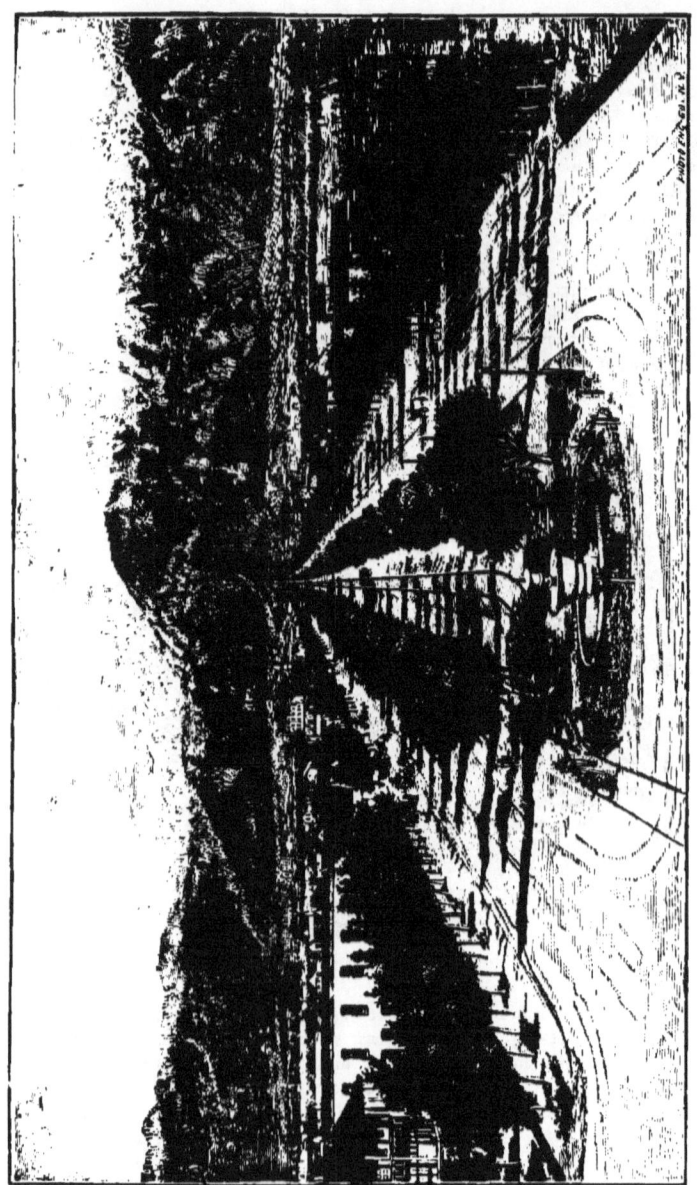

EUCLID AVENUE: ONTARIO, CAL.

a capacity of 500 kegs of nails a day, and will be one of the important industries of the San Diego region.

The Sweetwater Dam. This dam is one of the engineering wonders of this region, and an excursion to it is a most enjoyable experience. It is situated about six miles back of National City, and is reached by the National City & Otay Railroad. The dam, together with sixty-five miles of wrought iron pipe laid from the reservoir to National City, and to various points in that section for irrigation purposes, cost a total of $800,000. The dimensions of the dam are as follows: 46 feet in thickness at the base, 12 feet in thickness at the top, 75 feet in length at the base, 396 feet in length at the top. The reservoir is three miles long, three fourths of a mile wide, and covers 700 acres. When full it will hold six billion gallons of water, a quantity sufficient to irrigate 30,000 acres of land and supply a city of 50,000 people for one year, or irrigate 50,000 acres of land one year.

THE CLIMATE.
Summer the Year Round.

The Home of Health and Pleasure.

The climate of this region is a perpetual source of wonder to visitors. It is stating the simple, unquestionable fact to say that it has no equal among the health resorts of the world. From the compiled records of the U. S. Signal station here we extract the following: From 1876 to 1885, both years inclusive, covering a period of ten years, and embracing a period of 3,653 days, there were 3,533 days on which the mercury did not rise above 80°; and only 120 days in ten years in which the thermometer marked a higher temperature than 80°. During these ten years there were never more than two days in any one month in which the mercury rose as high as 85°, except June, 1877, four days, September, 1878, five days; June, 1879, two days; September, 1879, four days.

Returning to Los Angeles. The lovers of fine scenery, yachting, ocean bathing, salt sea fishing, outings among the hills, and those who delight in a summer which circles the entire year, will most reluctantly tear themselves away from the charms of San Diego. But one can't travel and stand still at the same time; so we take a night train northward on the same line we came in on, and sleep sweetly in one of Pullman's Palaces until we reach Colton. Here, after a good breakfast, we take the Southern Pacific road for Los Angeles, thus passing through new scenes from this point on to our destination. The first station reached after leaving Colton is

Cocamonga. This town is situated in the region made familiar to the public by the Cocamonga wine, the grapes here being noted for their fine quality. Slover Mountain is near Cocamonga, and is remarkable for containing quarries of onyx, lime, marble and cement. The "Mountain" is in reality only a moderate sized conical hill, but its rich deposits make it more valuable than a whole range of its big brothers. The marble is of the best quality, and can be quarried in great blocks, fifty feet long, if desired, and with a width of from five to six feet. The onyx is white, and is mined in large quantities for ornamental uses. Along the southern foot of Slover mountain flows the river Santa Ana.

Ontario is located on the main lines of the Southern Pacific and the Santa Fé Railways, the main depot being on the Southern Pacific, 38 miles from Los Angeles and 20 from Colton, while the Santa Fé runs two miles north, the station being North Ontario. From the Southern Pacific depot, the Chino Valley Nar-

CASCADE IN SAN ANTONIO CANON. ONTARIO, CAL.

SAN ANTONIO FALLS. ONTARIO, CAL.

row Gauge is built through the Chino Ranch, by the town of Chino, running three daily trains each way. The Southern Pacific and Santa Fé run also three trains daily each way, thus affording first-class railway facilities; besides which, the proprietor of the Chino Ranch & Chino Valley Railroad proposes to extend immediately his railroad to the sea, in the neighborhood of the new harbor of San Pedro. Ontario comprises some 12,000 acres, located on the mesa which slopes south gradually from the Sierra Madre Mountains to the Santa Ana River. It is in the west part of what is commonly known as the San Bernardino Valley, and occupies the highest point passed by rail or carriage road between Los Angeles and San Bernardino. The lands reach from the mountains around the San Antonio cañon to the Chino Ranch, a distance of about nine miles, and the Colony ranges in width from one to three miles. The altitude is a little less than 1,000 feet at the ranch line, and the grade is about 100 feet to the mile, increasing a little nearer the mountains, the mouth of the cañon being about 2,200 feet above sea level.

The scenery around Ontario is of the most striking and attractive character. To the northwest rise the Sierra Madre Mountains, while to the east towers the San Bernardino Range, and to the west slumbers the dreamy Pacific Ocean. As special landmarks in this striking scene are the four highest peaks of Southern California, namely: Mount San Bernardino and Old Grayback to the east, San Jacinto to the southeast, and Mt. San Antonio (Old Baldy) adjoining the Ontario tract on the north. Ontario occupies the elevated plateau between the San Bernardino Mountains and the ocean. The mountains being closely adjacent, and the sea being forty miles distant. The settler can choose his altitude from 900 to 2,500 feet, and by so doing find exactly the climate that is suited to his personal tastes On the higher slopes of Ontario we can see orange groves bearing fruit and flowers in delightful profusion, suggesting the breezes of "Araby the Blest," while half a dozen miles distant on the mountain peaks gleams the arctic snows. Nowhere in the world are summer and winter brought into closer juxtaposition. The zones of perpetual summer and never-ending winter are separated only by the San Antonio Cañon. Nor is it scenery alone which recommends the "Model Colony" of Ontario. Here are the best fruit lands in this country of fruit producing acres. Here the orange and the lime grow most perfectly and most abundantly; here deciduous fruits flourish, and here, in a word, is the fruit growers' paradise. It is alleged that orange groves at less than four years of age have produced, and frequently do produce, from $300 to $500 worth of fruit per acre. So great is the fertility, indeed, that three year old trees have been known to produce a full box of oranges each. But oranges are not the sole products of this wonderful soil. There are grown in great profusion the olive, peach, apricot, guava, prune, pear, apple, persimmon, plum, raisin and grape, and when one has mentioned these, he has only begun the list. For residence there can be no pleasanter place than Ontario, and for horticulture and arboriculture surely no place can claim precedence. For further information concerning this place, the reader is referred to the Addenda.

POMONA.

Health and Pleasure Resort.

A Fruit Growing Paradise.

One of the prettiest towns in the San Bernardino Valley is Pomona, in the eastern part of Los Angeles county, thirty miles from the city of Los Angeles and thirty miles from the Pacific Ocean northward and fifty miles eastward. The Sierra Madre range of mountains—average elevation of 9,000 feet above the sea, with snow-capped peaks—are distant six miles north, and Mt. San Bernardino (height 11,000 feet) and Mt. San Jacinto (about the same height) forty and fifty miles eastward. The lower range, called the San Jose Hills, midway between the Sierra Madre Range and the ocean, terminates at the city, and the great valley widens at this point to twenty-five and thirty miles.

Thus these high mountain ranges protect this valley equally from harsh sea winds and the unpleasant dry winds and sand storms of the desert. The altitude of the city is 860 feet above the sea, the valley rising gradually to 2,000 feet at the foot of the mountains. The immediate locality bears a similar relation to the mountains and the ocean as the celebrated health resorts of Mentone and Nice.

A ride through the streets of the city, or along the many roads traversing the country in every direction, will disclose many fine residences; also cosy, comfortable homes. Houses, which are neither large nor costly, show the refinement of true comfort and adaptation to the wants of the owners. The mild, open winters, and consequent freedom from cold, do not require as expensive houses as in eastern and northern climates; therefore the house is open, cheery and home-like in its appointments, many with broad verandas for the open-air life of the occupants during most of the days of the year; and yet the individuality of the owner is as plainly seen in the architecture and plan of the modest home as the more pretentious building of the city or in older communities; for these quiet homes are surrounded by groves of trees, many of them evergreen—rows of vines extending almost as far as the eye can reach—with roses and flowers from the roadside to and surrounding the house, the whole deeply impressing the visitor with the air of home-like comfort and cheerfulness everywhere prevailing. The town is amply supplied with water, this precious fluid being obtained from three sources, namely, San Antonio Cañon, numerous *cienegas* which encircle the valley, and which are fed by subterranean streams from the high mountains and artesian wells. There are in this valley some of the finest flowing wells upon the continent, some of which have given an undiminished flow for nearly ten years. There are now flowing in the Pomona Valley sixty-seven wells, fifty-two of which are owned by the Pomona Land and Water Company, who are extending their works at different points and increasing the number. These waters are alike free from alkaline, saline or mineral taint, and deliciously cool and invigorating. The right to use water for irrigation is sold with the land, so that there need be no fear of a lack of this necessity upon the part of those who settle here. Additional information concerning this place will be found in the Addenda.

As Pomona is directly suggestive of the subject of fruit, and as fruit culture is *the* great industry of Southern California, this is an appropriate place to introduce a few statistics on the subject. We condense the following facts from reliable documents:

The cost of raw land may safely be placed at $150 per acre, which is about an average, according to location. In the following estimates for a vineyard, the Zinfandel and Berger grapes have been taken, varieties which have been tested here, and which have proven highly satisfactory:

Ten acres of land, @ $150 per acre	$1,500
Two plowings, leveling, etc., @ $5 per acre	50
Cost of cutting for 10 acres	50
Planting, @ $5 per acre	50
Care for two years, @ $15 per acre	300
Total cost till brought to bearing	$1,950
Crop third year, 5 tons to acre, @ $20 per ton	$1,000
Crop fourth year, 7 tons to acre, @ $20 per ton	1,400
Crop fifth year, 10 tons per acre, @ $20 per ton	2,000
Total for three years	$4,400
Deduct cost of care for third year, fourth and fifth years, @ $15 per acre	$ 450
Cost of land, vineyard, etc.	1,950
Interest 2 years, @ 10 per cent	390
	$2,790
Net profit, five years	$1,610

This estimate supposes that all the work is hired. If a man is not afraid to take hold and do most of the work himself, which he can easily do, the expense account would be materially smaller.

Prunes promise to be a most profitable fruit, and have proven themselves at home in this valley.

Cost of ten acres of land, as given above	$1,500
Plowing, etc.	50
Cost of prune trees for ten acres	180
Planting	50
Care for 3 years, @ $15 per acre per year	450
Cost of 10-acre prune orchard to time of bearing	$2,230

ORANGE ORCHARD: POMONA, CAL.

The fourth year the account stands about thus:

Crop, 100 lbs. per tree and 108 trees per acre, 108,000 lbs.,
 @ 2 cts. per lb..$2,160
Fifth year, 150 lbs. per tree, 162,000 lbs., @ 2 cts 3,240
                                                              ———
                                                              $5,400
Deduct cost of orchard $2,230
Interest on $2,230 for 3 years, @ 10 per cent............ 669
                                                              ———
                                                              $2,899
 Net profits for five years..............................$2,501

Other deciduous trees will show about the same result. As has been stated, the expense account can be largely decreased if a man is willing to take hold and work. One horse will do all the cultivating; thirty dollars will buy all the implements needed, and there need be no expense for hired help until the grapes or fruit are to be gathered, and enough potatoes and other vegetables can be raised on the land to furnish a living for the first few years.

Beyond Pomona are a number of small stations possessing all the requisites of climate, soil and scenery to become thriving towns; which, doubtless, will be the outcome in a few years. At present, however, they possess only a statistical value to the tourist. These stations occur in the following order: Spadro, Lemon, Puenta, Monte and Savanna.

San Gabriel. This is the site of the famous Mission of San Gabriel, or, to give it the full honors of its stately Spanish title, "El Mission de San Gabriel Arcangel." The Mission was founded September 8, 1771, and was moved from the original site to its present position in 1775. The mission church is plainly to be seen from the car windows to our right, just after the station has been passed, and is a most interesting relic of what in the new world may be called antiquity, having been erected, in 1804, of material imported from the mother country, Spain.

Beyond San Gabriel are the suburban stations to Los Angeles, of Alhambra, Shorb and Aurant. The handsome suburban villas which dot the landscape on each hand rapidly increase in number as the city is approached, and soon we are rolling along between continuous rows of houses, and finally come to a stop at the Southern Pacific Railroad's depot, in Los Angeles.

SANTA MONICA.
The
Long Branch
of
the Pacific.
A charming Sea Shore
Watering Place.

The trip from Los Angeles to Santa Monica, one of the famous bathing resorts of the Pacific coast, is not only justified by what one finds at the end of his journey, but also by the pleasures enjoyed *en route*. The Los Angeles & Independence Railroad runs four trains to the beach each day—a distance of sixteen miles; and on Sunday the exodus to this famed seaside resort is something extraordinary. For three or four miles after leaving the station, we pass through suburbs of Los Angeles. Handsome villa residences, surrounded by beautiful and most attractive grounds, are to be seen on every side. At last, reaching the open country, we pass through a constant succession of vineyards and fruit orchards, until the near presence of the ocean is made known by refreshing saline breezes and the occurrence of sand dunes and salt marshes. The train stops at a handsome depot, beyond which extends a large, well-kept and beautiful park. It is difficult for one accustomed to the varying seasons of the lands across the mountains to comprehend the fact that this beautiful park, with its luxuriance of sub-tropical vegetation, its affluence of delicate and vari-tinted flowers, is never less verdant, less brilliant or less attractive than it is now. It is not easy to grasp the fact that all the year round, equally as comfortably on the first of January as on the first of June, one can sport among

SANTA CRUZ.

OCEAN SCULPTURE — SANTA MONICA.

the combing billows that come rolling in across the blue, serene Pacific. The attractions of Santa Monica are manifold, — beach-driving, surf-bathing, fishing, boating, yachting, are the sea-ward delights; while on the shore are all the charms which nature has so opulently spread for the pleasure of those who visit this favored spot, together with all the ingenious devices invented by man for amusement and relaxation. Of course, it goes without saying that there is a magnificent beach hotel, whose broad verandas face the sea, and whose appointments are complete in all respects; also, of course, there are bath-houses of ample accommodations.

There are many points of scenic interest within easy-driving distance of Santa Monica. One of the most charming is that to Santa Monica cañon and Manville Glen, a spot made cool and inviting by ancient forest trees and a rippling brook, all embraced by rugged mountain surroundings. This is a favorite camping ground, where pleasure and health seekers pitch their tents and spend months in the calm enjoyment of this sylvan retreat. Santa Monica is a great health resort, and experience has proved its excellence in this regard. It possesses, the year round, one of the most enjoyable and healthy climates in the world, being from ten to fifteen degrees cooler than Los Angeles and the interior country in summer, and warmer in winter. There is a magnificent driving beach stretching away for fifteen miles, good sea fishing, an abundance of water fowl in the neighboring lagoons, and game in the mountains a few miles distant. There is a capacious, deep-water roadstead, with good anchorage, where vessels may lie in safety the greater portion of the year. The climate of Santa Monica is worthy of somewhat extended notice. In a general way we can sum up the climatic conditions of the Southern California coast as follows: So far as the amount of rainfall is concerned throughout Southern California, the rainy season simply signifies that during that period, exclusively, not exceeding 18 inches may fall. The average annual rainfall at San Diego is only 10.43 inches. Following up the coast to San Francisco, it increases at the rate of about 2 inches for every 100 miles. Santa Monica receives about 13 inches, Santa Barbara 15 inches, Monterey 17 inches, and San Francisco 21 inches. The Coast Range of mountains, rising to an elevation of from 2,000 to 4,000 feet, robs the ocean rain-freighted clouds of all their precious burden before reaching the interior plains and valleys. At Fort Yuma, on the Colorado River and Desert, the mean annual rainfall is only 2.54 inches; among the little valleys extending from San Diego to the San Jacinto Mountains, from 7 to 9 inches; in the valley of San Bernardino, and at Colton, Riverside and Cocamongo, 10 inches; advancing toward the coast, Spadra and El Monte receive about 11 inches; and Los Angeles, situated 20 miles from the ocean, about 14 inches. Crossing the San Bernardino Mountains to the Mojave Plains, the yearly rainfall is only from 3 to 4 inches, and from thence up the San Joaquin Valley as far as Goshen, in latitude 36 degrees, it ranges from 3 to 6 inches; from thence, northward, it increases to 15.10 at Stockton and 18.23 at Sacramento. Taking it all in all, Santa Monica is a place of great interest. We have said nothing about the town so far, but must not neglect to state that there *is* a town, and a very pretty one at that. It is situated on the level mesa, which stretches back landward from the brink of the natural sea wall, from whose foot extends the level beach outward to the ocean rim. The residences are tasteful, many of them elegant, the business blocks substantial, and every element of comfort and convenience for the health or pleasure seeker can be found here. For further information concerning this resort, the reader is referred to the Addenda.

Long Beach. We have already described the greater portion of the trip from Los Angeles to Long Beach in that portion of this book devoted to the journey from Los Angeles to San Pedro. We follow the same line in our excursion to the Beach as far as the Junction, at which point our train takes the line to the left, and rolling along through a level country, encroached upon here and there by the salt marshes of the ocean, but passing many fertile and attractive spots, soon reaches Long Beach, the goal of our journey. This popular resort is only twenty five miles distant from Los Angeles, and can be reached in an hour's ride from the city. A fine hotel has been built here, which overlooks the ocean and the beach. Surf-bathing may be enjoyed here the year round, and the accommodations are complete in every respect. The beach itself is one of the greatest attractions of the place. The sands are left hard and compact by the retiring tide, and the drive along the margin of the ocean is undoubtedly the finest to be found anywhere on the California coast. Long Beach has a wharf which extends a distance of 750 feet in the ocean, reaching water deep enough to float vessels of the heaviest tonnage by its side. Long Beach has already become a resort of great popularity, and the excellence of its beach, its attractive scenery and fine hotel combine to render this popularity greater every day. For further information concerning this resort, the reader is referred to the Addenda.

CALIFORNIA'S MAMMOTH GRAPE VINE.

SAUNTERINGS AROUND SAN FRANCISCO.

FTER enjoying the delights of Southern California, the tourist can return to San Francisco from Los Angeles over the same route by which the southward journey was made, or he can take steamer at Santa Barbara or San Pedro, and have the pleasure of a delightful coast voyage. Having once more established headquarters in the metropolis, he will be ready to make excursions to the points of interest adjacent to the city.

San Francisco to Monterey. It was a bright, genial California day, when we took the cars of the Northern Division of the Southern Pacific Railroad, at the station opposite the immense brick building at the corner of Fourth and Townsend streets, in which are the general offices of this great railroad company. We were bound for Monterey, famous for its bathing and its Rosamond's Bower—the world-renowned Hotel del Monte. Our course is southward through the city for a distance of four miles. Two miles from the station are the machine shops of the railroad company. Valencia street station is reached in another mile; here the cable line through the center of the city to Oakland Pier crosses the track. Beyond this station the suburbs of the city are entered. On the right, occupying an elevated position, is the Industrial School building. Bernal is passed, and numbers of market-gardens, with an intricate and interesting system of terraces and irrigating ditches, pipes and flumes. Holy Cross Cemetery is seen to our left, then Coloma, Ocean View and Baden come next and then we approach quite near the shore of San Francisco Bay, reaching this point by means of a sharply descending grade.

San Bruno. Here are the rifle ranges of the shooting clubs, situated on the shore of the bay; and here also is a large hotel, a popular resort for the sportsmen who congregate at this place. (Population, 50. Distance from San Francisco, 14 miles. Elevation, 15 feet.)

Millbrae. This is the station for the country-seat of Mr. D. O. Mills, president of the Bank of California, and his palatial residence can be seen about half a mile distant to the right, characterized by two lofty towers. Just beyond the station is the Millbrae Dairy, with a multitude of buildings showing the great extent of this enterprise. Handsome residences can be seen on the right, the left side being next the bay, and given over to meadow lands and cultivated fields, diversified by occasional groves. (Population, 100. Distance from San Francisco, 17 miles. Elevation, 8 feet.)

Fair Oaks. This station is the site of most attractive groves of live oaks, from which it takes its name. (Distance from San Francisco, 31 miles.)

San Mateo. Surrounding San Mateo are a number of the most elegant country-seats in California. Wealth has concentrated its forces here, and everything that money can do, when employed unstintedly and intelligently, has been done to beautify the scene. The art of the landscape gardener has here been exercised to its fullest extent, and the grounds which surround these palaces of San Franciscan millionaires are bewildering visions of arboreal and floral beauty. As we advance after leaving the station, the race track is passed on the right, also the Young Ladies Seminary. The bay is on the left. Groves of oak, eucalyptus

OLD MISSION CHURCH — MONTEREY.

trees and endless orchards stretch away to the right as far as vision can reach. Four miles beyond San Mateo is Belmont, the station for the country-seat of the late banker king, John Ralston, which is one of the most noted country residences near San Francisco, and during the life of its owner it was the scene of a most generous and lavish hospitality. The statistics of San Mateo are as follows: (Population, 950. Distance from San Francisco, 21 miles. Elevation 22 feet.)

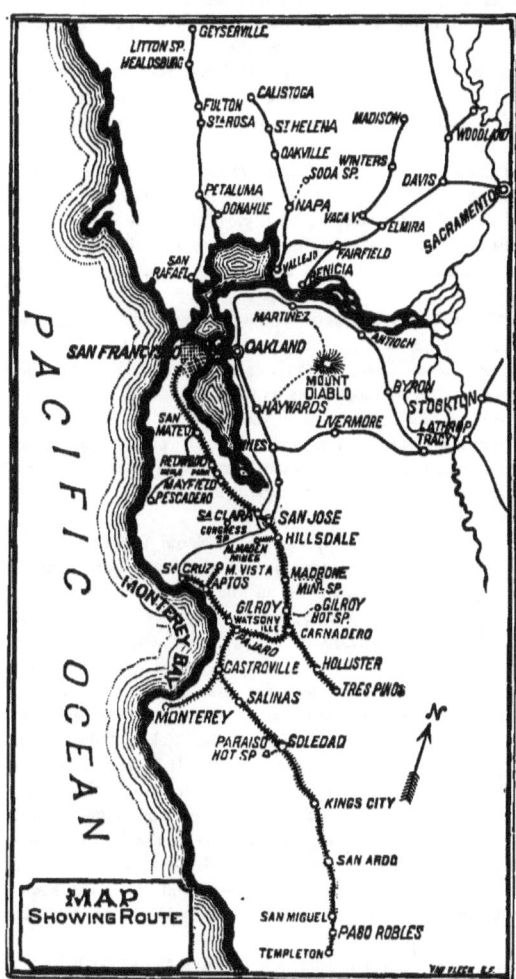

MAP SHOWING ROUTE

Redwood. This town is the county-seat of San Mateo county, and derives its name from the proximity of extensive redwood forests to the westward. Great quantities of redwood lumber, firewood and bark are shipped from this station. Artesian wells furnish water for the town, which is well built and supplied with all of the modern conveniences. The tributary industries, in addition to lumbering, are fruit raising and agriculture. (Population, 1,500. Distance from San Francisco, 28 miles. Elevation, 9 feet.)

Menlo Park is noted as the residence of a large number of San Francisco's most wealthy business men and gentlemen of leisure. It is a bower of beauty in the heart of umbrageous groves, made still more lovely and attractive by flowers of every hue and a generous abundance of ornamental shrubs and trailing vines. It goes without saying that all that the genius of the architect can devise has been done to make the country residences equal in beauty with their surroundings. (Population, 400. Distance from San Francisco, 32 miles. Elevation, 64 feet.)

Beyond Menlo Park, on our right, is the stock farm of Senator Leland Stanford, president of the Southern Pacific Railroad Company. The farm embraces five hundred acres, and here have been bred blooded stock, the finest of any on the Pacific coast.

Leland Stanford, Jr., University. The site of this university, which has an endowment of $20,000,000, and is a monument of parental affection, is Palo Alto, near Mayfield, two miles beyond Menlo Park. There are 4,291 acres of land in the grounds belonging to the university estate.

Alviso. This station is at the head of San Francisco Bay, and from this point great quantities of fruit, especially of the smaller varieties, are shipped by boat to San Francisco. (Population, 110. Distance from San Francisco, 38 miles. Elevation, 8 feet.)

Santa Clara was founded by the Jesuits in 1774, and has for its site a most beautiful region, being near the centre of the fertile Santa Clara Valley. The climate is noted for its healthfulness and equability. This valley is one of the best wheat regions in the state, and is also noted for the abundance and fine quality of its fruit. Santa Clara and San Jose are twin cities, being only three miles apart.

From San Jose, the Alameda, a broad and famous avenue lined with ancient willows, leads to the old town of Santa Clara, four miles distant. The Mission of Santa Clara was founded by Father Pena, in 1777, and the old adobe walls are still crumbling away. Twelve miles by stage takes us to the famous New Almedan quicksilver mines, which furnish half the quicksilver the world produces, and gives employment to several hundred miners. The mountains are picturesque, easily reached, abound in trout and game, and contain many health and pleasure resorts, besides presenting every attraction to camping parties. (Population, 3,000. Distance from San Francisco, 44 miles. Elevation, 75 feet.)

SAN JOSE.
Metropolis of Santa Clara Valley.

The Garden City of the Pacific Coast.

The metropolis of the Santa Clara Valley is San Jose, the county seat of Santa Clara county, and the Garden City of the Pacific Coast. It is a progressive and rapidly growing city, with a population of 25,000. It is fifty miles distant from San Francisco, with which it has rapid and convenient communication by three lines of railroad, operated by the Southern Pacific Company, giving trains either way at all times of day, tickets being interchangeable on all the routes. It is also convenient to the most charming seaside resorts in the world, Santa Cruz and Monterey, and two special excursion trains are run to these places weekly. The beautiful surroundings and delightful climate of the valley already briefly alluded to, its many elegant and costly homes, its shaded streets and avenues literally embowered in trees of perpetual verdure, and the many social and other advantages which wealth and culture have bestowed, make San Jose the place for an ideal home.

There are twenty three churches in San Jose, and its educational facilities are unsurpassed. The five public schools are not excelled in the state, and a high school with an advanced curriculum crowns the system. There are here, besides, some of the best known educational institutions on the Pacific Coast. The University of the Pacific is a Methodist institution of high rank; the State Normal School and the College of Notre Dame are centrally located, and at Santa Clara, three miles distant, is the Santa Clara College, an institution located on the site of the old Santa Clara Mission. At Palo Alto, a few miles to the northwest, are being erected the buildings of the Leland Stanford, Jr., University, one of the grandest educational institutions of the world, having an endowment of $20,000,000. There are also, in and about San Jose, several other minor educational institutions. Besides the vast and only partial developed resources of the Santa Clara and three small but fertile tributary valleys, San Jose has many industries as a basis for her

prosperity. Recent experiment has shown that the Santa Clara Valley is one of the most favorable regions in the world for the propagation of the silk worm, and the infant silk industry is already represented by a manufactory of dress silks. There are four large fruit canneries, three glove factories, two flouring mills, a large woolen mill, and a great number of other industries. Four miles away are the great Lick paper mills. (Population, 25,000. Distance from San Francisco, 50 miles. Elevation, 86 feet.)

ON THE BEACH AT SANTA CRUZ

The Lick Observatory. Throughout this region are many points and features of interest. First in importance is the great Lick Observatory, whose dome glistens in the sunlight on the top of Mount Hamilton, twenty-six miles away, and at an altitude of 4,443 feet. Here is now in place the most powerful telescope in the world. For this observatory the late James Lick bequeathed $700,000, and the property now belongs to the University of California. Daily stages run to the summit over a magnificent winding road, which cost Santa Clara county $100,000. From the summit, on a clear day, the view is one of indescribable beauty and grandeur. The great dome of the observatory can be plainly seen to the left, from the windows of the train, after San Jose has been left behind. Two delightful side trips from San Jose are those to Los Gatos and Santa Cruz.

Los Gatos. Nine miles from San Jose, on the direct narrow gauge line to Santa Cruz, lies Los Gatos, a thriving town of 2,000 inhabitants, which nestles amid picturesque surroundings, on the eastern slope of the Santa Cruz Mountains. It is within the Santa Clara Valley, partaking of all the material and climatic blessings of that lovely region; but it also lies within the thermal, or warm, belt, and so enjoys an added advantage and attraction. This thermal belt is an interesting phenomenon, and is observed in all the foot hills of the bay region. It is due to the fact that when the cooler airs of night flow into the broad valley below, sometimes lowering the temperature until frost is formed, the warm air rises and rests at a higher altitude, preserving in a wide strip of country along the mountain sides a higher temperature at night and a more equable climate than is found in the valley below. About Los Gatos, the strip of country so affected is six miles in width. The change is quickly noted in a drive from San Jose to Los Gatos in the cool of the evening. This condition gives to Los Gatos a truly Arcadian air that is a perpetual delight to the visitor or resident. It is this which makes the successful cultivation of the orange, lemon and other citrus fruits possible here. The soil here is as rich and fertile as in any portion of the valley, and this with the perfect climate, pure and balmy airs, and the ever present beauty and abundance which has followed the efforts of labor and capital, make of this particular region a veritable Eden.

The country about Los Gatos is noted for its fruit, and the whole slope is covered with profitable vineyards and orchards, wherein plums, peaches, prunes, apricots, pears, apples, olives, figs, cherries, oranges, lemons and other fruits attain a rare perfection. English walnuts are extensively raised, and the largest almond orchard in the world is located here. Owing to its proximity to San Francisco, and its unequaled attractions as a place of residence, Los Gatos has become the suburban residence place of a number of San Francisco men of wealth, and a special suburban train is run to the town. (Population, 2,000. Distance from San Francisco, 55 miles. Elevation, 400 feet.)

SANTA CRUZ.

The Newport of the Pacific Coast.

Health and Pleasure Resort.

One of the loveliest cities of California is that of Santa Cruz. Occupying a charming site on the seashore at the north end of the crescent-shaped bay of Monterey, it is at the mouth of the San Lorenzo river, and recedes from a beautiful beach, extending to a broad plateau and two terraces rising above it, surrounded by protecting hills.

Santa Cruz is the most popular and fashionable seaside resort in the state, and is termed the Newport of the Pacific Coast. During the summer season, people flock to this beautiful city by the thousands, especially from

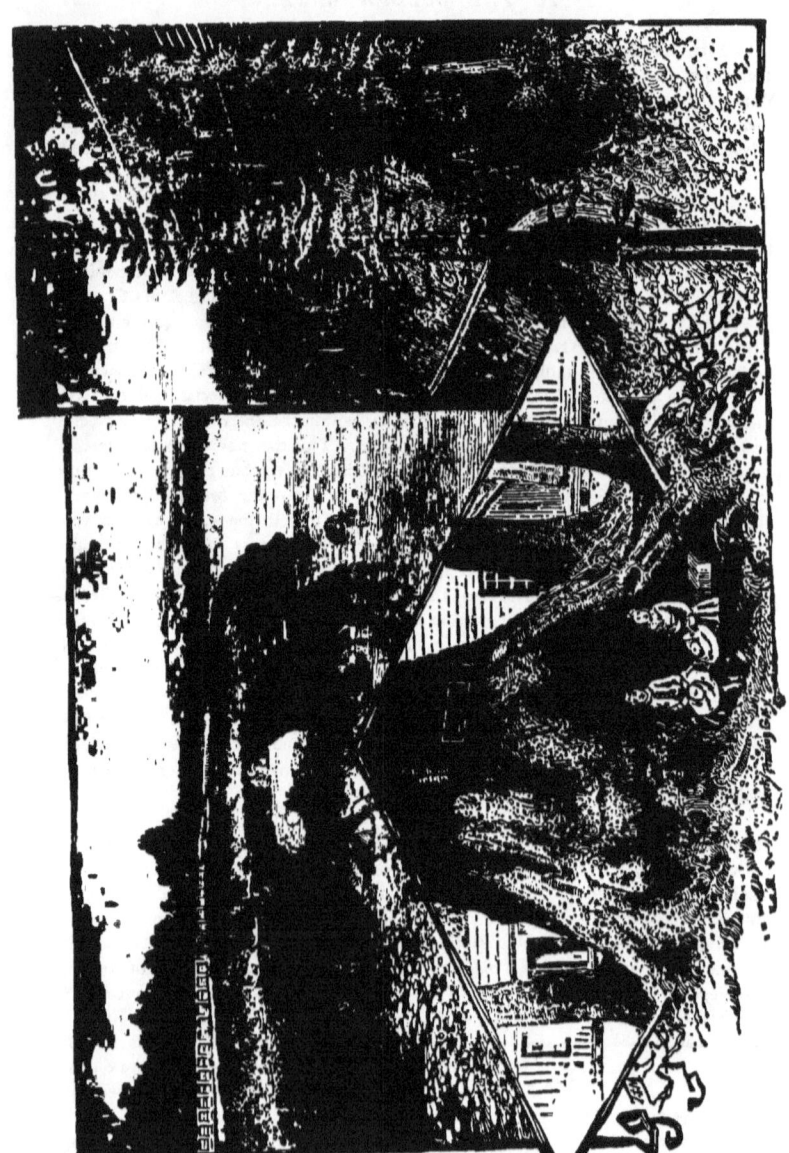

"GENL. FREMONT" BIG TREE, AND BEACH VIEW.

the cities; and during the "season," the population is increased to ten or twelve thousand. Its bathing is its pride and its glory, and with its lovely beach of clean, white sand, its fine bath houses, and its safe and delightful waters, its attractions in this line are unequaled. While the crowds are greatest during the vacation season, between May and September, the bathing is delightful the whole year round. Excursion trains are run to this place from San Francisco every Saturday and Sunday.

The place itself has all the features of a modern progressive city, among which may be mentioned electric lights, gas, fine water system, fire department, street cars, free library, telephones, three daily and two weekly papers, two banks, handsome public buildings, fine schools and numerous churches. It has magnificent streets, many being paved with bituminous rock, and its sidewalks are exceptionally good. Its great number of cosy and attractive homes are among its chief features.

The climate of Santa Cruz and surrounding country is unsurpassed by that of any other part of the state. It is remarkably equable, the average temperature for winter being 52 degrees, and for summer 62 degrees. The difference between the extremes of the year is always small. Epidemics are unknown, and health and vigor is imparted to invalids, as well as to the strong. Rare roses, and other blooms which in the East are hot-house plants, are culled from gardens every week in the year.

In the mountains and valleys about Santa Cruz, there is an almost endless array of attractions. The mountains are exceedingly picturesque, abound in trout and game, offer great attractions to the camper, are full of beautiful cañons and nooks, and the views to be obtained are of surpassing beauty and grandeur. Ben Lomond is the highest peak, and is reached by four delightful routes. Five miles from Santa Cruz is a famous group of redwoods, known as the "Big Trees." The largest is 300 feet in height and 60 feet in circumference. There are several valleys of great beauty and fertility, the most noted being the Pajaro Valley. In the valleys and on the mountain slopes fruits of all kinds are extensively grown, with as great success as anywhere in the state, and general farming is easy and profitable. The dairying interests are extensive, and the forests are still large and dense. The county is the second in the state in manufactures. This is a region of flowers and perpetual summer, with every attraction for residence, and offering great inducements for investments. (Population, 7,000. Distance from San Francisco, 80 miles. Elevation, 15 feet.)

Castroville. Resuming our journey at San Jose for Monterey, we pass through an interesting and fertile country until Castroville is reached. From this point a branch extends down to Monterey, the main line running to Templeton. Around Castroville is one of the greatest wheat growing regions of California. The ordinary yield is from 40 to 50 bushels to the acre, though as high as 102 bushels have been grown here—the largest yield on record. (Population, 600. Distance from San Francisco, 110 miles. Elevation, 17 feet.)

From Castroville we follow the curving shore of the Bay of Monterey; at times within a short distance of the shore, and at others somewhat farther inland. Sand dunes and salt marshes testify to the nearness of the ocean.

Del Monte. In the heart of a lovely grove the train stops at a tasteful rustic pavilion, which is the station. Broad, graveled roads sweep up to the station in graceful curves, and here stand waiting richly appointed four-horse carriages, in which guests for the famous Hotel del Monte are conveyed to their destination. Glimpses of the hotel to the left can be caught through the interstices of the trees; while vines and shrubs and flowers grow everywhere in studied and

ON THE BALCONY — HOTEL DEL MONTE.

...istic confusion. The drive to the hotel along the broad, tree lined avenues, shaded by immemorial and stately live oaks, through which vistas of sylvan beauty can be seen, gives one a foretaste of the charms of this one of the most charming places in the world.

> **THE HOTEL DEL MONTE.**
> A Palace of Delight.
>
> The Queen of American Watering Places.

The hotel is first seen through a vista of trees. and, in its beautiful embowerment of foliage and flowers, resembles some rich private home in the midst of a broad park. This impression is heightened when the broader extent of avenues, lawns and flower-bordered walks come into view. The gardener's art has turned many acres into a choice conservatory, where the richest flowers blossom in profusion. Here and there are swings, croquet grounds, an archery, lawn-tennis courts, and bins of fine beach sand—the latter being intended for the use and amusement of the children, who can not await the bathing hour for the daily visit to the beach. The use of all these, as well as of the ladies' billiard saloon, is free to guests. In all directions there are seats for loungers. Through a vista formed by the umbrageous oaks and pines, the huge, bulbous forms of a varied family of cacti are seen. In another place is a bewildering maze. Everywhere flowers and rare plants abound, and every avenue and pathway is bordered by intricate floral devices. In any direction the eye may turn are fresh visions of beauty. In the fall of 1883 a great improvement was consummated in the introduction of an abundant supply of pure, soft water from the Carmel river. Extensive water works were constructed at an expense of over half a million dollars. The supply not only meets every requirement of the hotel, but also feeds the great fountain in the lake. The Del Monte Bathing Pavilion is situated on the beach, about eight minutes' walk from the hotel, and is one of the largest and most complete establishments of the kind in the world. It is seventy feet wide by one hundred and seventy long. There are four tanks of about thirty-six feet wide by fifty feet long. The water in t ese tanks ranges in temperature from cold up to warm, and the bather can take his choice. The heating is done by steam, and the water is daily changed. The pavilion contains two hundred and ten dressing-rooms, one half of which is set apart for the use of ladies. Each of the latter is fitted up with a fresh water shower bath, while on the gentlemen's side fourteen shower baths serve for all. The pavilion and everything connected with it is kept scrupulously clean, and always presents a pleasing appearance. When filled with bathers and spectators, it presents a spectacle which, in point of animation and interest, would be hard to surpass. Outside of this pavilion is a beautiful sandy beach, on which surf-bathing may be indulged. An adjunct of the Hotel del Monte is its 18 mile drive, over a splendidly-kept macadamized road, by way of Monterey, Pacific Grove, Cypress Grove, Carmel Bay, and the old Mission Church. The reader will remember the sensation which was created several years ago by the burning of the Hotel del Monte. From its ruins there has arisen a new Del Monte—larger, more beautiful and complete than the old one. The new Del Monte is in its main front and general style of architecture an exact copy of the old Del Monte, which was universally pronounced, by thousands of famous visitors from all countries, to have been the most graceful and elegant building of its class in the world. The new building, by increasing and extending its annexes, has nearly double the accommodations of the old one. These annexes are connected together by two arcades of glass and iron, three stories in height, which virtually makes the two annexes one.

INSIDE AND OUTSIDE HEADERS—DEL MONTE.

Being fully inclosed, and yet light as day, the guest experiences no inconvenience of any kind in walking through them; on the contrary, the two arcades make delightful little promenading places. The dining-room is 162 by 66 feet, nearly double the size of the old one, and will comfortably seat 500 people at once. The park grounds surrounding the Del Monte have no equal on this coast, and it is a mere question of time when they will have no superior anywhere. Nature endowed them with prodigal liberality, and the owners are supplementing nature's efforts with an equally prodigal expenditure of art. Croquet plats, an archery ground swings, lawn tennis grounds, choice flowers, shrubs, trees, beautiful walks, and in short, everything which an experienced landscape gardener's artistic eye can suggest, is being done for the improvement of this favored spot.

Monterey. This quaint and romantic old town, the capital of California when the territory was acquired by the United States, and the place where Fremont first raised the stars and stripes and took formal possession of the country, is one of the most interesting places to visit in California. Monterey is situated on the lovely bay of the same name, 125 miles from San Francisco by the Southern Pacific Railroad, and can be reached in 3½ hours by taking the fast Monterey train, leaving the city at 3.30 P. M. This is the fastest train on the Pacific Coast and one of the most elegant in equipment in the world. There is probably no place upon the Pacific Coast so replete with natural charms as Monterey. Its exquisite beauty and variety of scenery is diversified with ocean, bay, lake and streamlet; mountain, hill and valley; and groves of oak, cypress, spruce, pine and other trees. The mountain views are very beautiful, particularly the Gabilan and Santa Cruz spurs. The Bay of Monterey is a magnificent sheet of water, and is twenty-eight miles from point to point. It is delightfully adapted to boating and yachting; and many kinds of fish may be taken at all seasons of the year. For bathing purposes the beach is all that could be desired—one long, bold sweep of wide, gently sloping, clean, white sands — the very perfection of a bathing beach; and so safe that children may play and bathe upon it with entire security. There are also great varieties of sea-mosses, shells, pebbles and agates, scattered here and there along the rim of the bay, fringed, as it is at all times, with the creamy ripple of the surf. (Population, 2,300. Distance from San Francisco, 125 miles. Elevation, 5 feet.)

Pacific Grove, a short distance from Monterey, is to the Pacific Coast what Nantucket, Martha's Vineyard and Ocean Grove, are to the Atlantic sea-side resorts, except that the Pacific Grove retreat has as equable a temperature as Monterey itself, and is kept open all the year round. It is delightfully situated on the beautiful Bay of Monterey, less than two miles from the old town, and in loveliness of location cannot be excelled, its graceful pines extending to the water's edge.

YOSEMITE VALLEY.

TO THE YOSEMITE.

 one who visits San Francisco can afford to return home without seeing nature's great temple of wonders—the Yosemite. The way thither has been greatly smoothed by the Southern Pacific Railroad, and each succeeding year sees improvements in this direction. What was formerly an undertaking of considerable magnitude and difficulty, has now become an easy journey, and one fraught with pleasure in the taking It is only a vacation jaunt, requiring four days to make the round trip. The valley is 259 miles from San Francisco, 178 miles to Berende, on the route already described in the trip to Los Angeles, thence 21 miles by rail to Raymond, and 60 miles by stage to the valley. It is now all rail to the foot-hills of the Sierra Nevada Mountains, where the traveler is transferred to the most approved pattern of stages (or carriages, really), and is delightfully whirled up into the Land of Wonders over an excellent road, through giant timber, across ice-cold rivulets, and past cataracts which send their spray into the sunlight, embellished with the colors of the rainbow. Mr. Ben. C. Trueman, the veteran traveler and writer of the Pacific Coast, speaks as follows concerning this wonderland: "Some few years ago we visited the Yosemite in company with a gentleman who had traveled largely, and who had written much of the scenic attractions of Europe, Asia and America, and who exclaimed, as we reached 'Inspiration Point': 'My God! self-convicted as a spendthrift in words, the only terms applicable to this spot I have wasted on minor scenes.' And it was, unfortunately, true, that language failed to give adequate utterance to the emotion of my friend upon that occasion, and his hitherto facile pen failed to perform its functions with its characteristic felicity and brilliancy. This has been the case with many, however, if not with all others; and, thus, the pre-eminent grandeur and magnificence of the Yosemite remains, after all, untold. Indeed, its charms must really be seen and felt; for it is an absolute fact, that neither pencil nor brush, nor photographic process, can give them faithful protraiture."

YOSEMITE.
A Valley of Wonders.
The
Climax of Grandeur
and Beauty.

The Yosemite Valley is about 150 miles, in an almost easterly direction, from San Francisco and nearly midway of the state, between the northern and southern boundaries; it was for many years the rendezvous, or permanent abiding place, of hostile Indians, who had a legend for every point of interest, whether of water or rock. The place was first seen in 1850 by a number of white men, who had formed themselves into a military company to punish or compel peace with bands of murderous Indians; it was taken possession of in March, 1851, by an expedition under the command of Captain Boling, which invaded the aboriginal stronghold, killed several of its defenders, and either stampeded or compelled peace with the rest. The valley is some 15 miles long by about one-third of that distance in width, and is undoubtedly the most wonderful combination of chasm and dome, cliff and cañon mountain and valley, river and waterfall, cataract and streamlet, winter and summer,

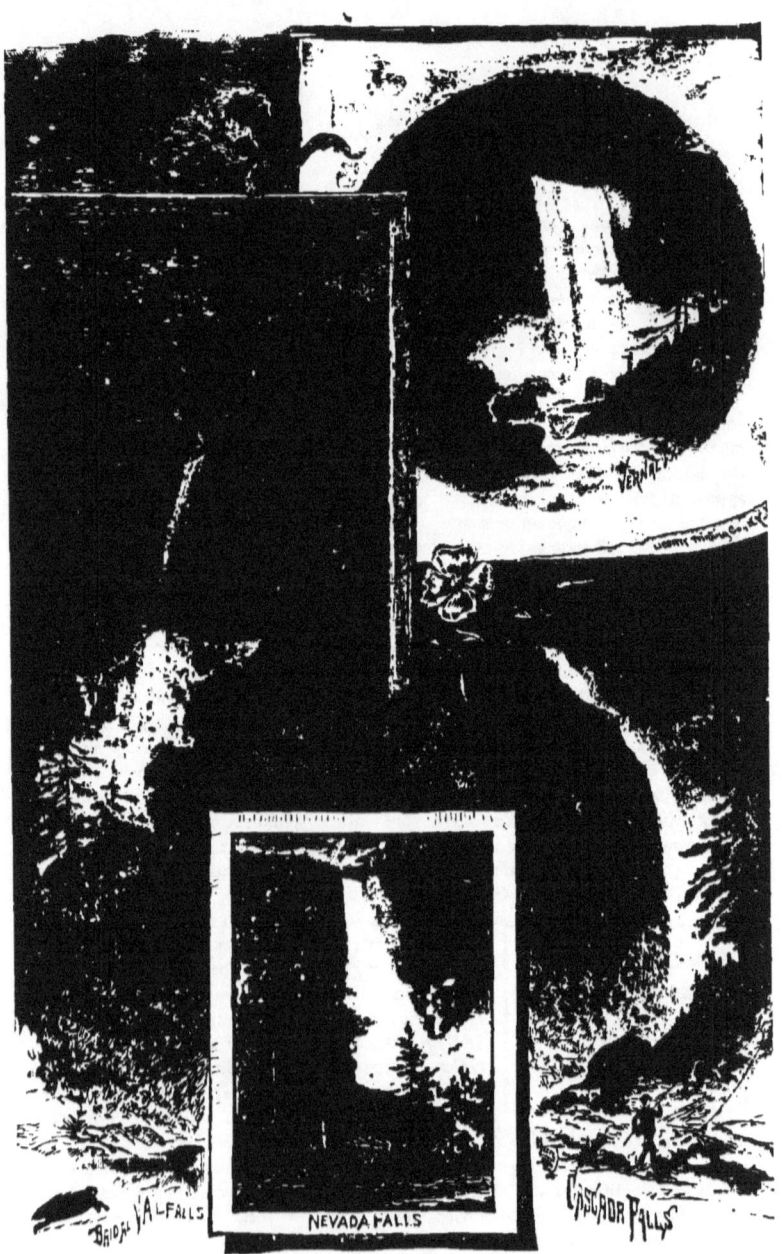

FALLS OF THE YOSEMITE

and sunshine and shadow, to be seen in the world—especially within a radius of eight or ten miles. Among the most noted and majestic elevations, which rise right up vertically, many of these seeming like hewn rock, are: El Capitan, 3,300 feet above the floor of the valley; Cathedral Rock, 2,660 feet above the valley; Three Brothers, 3,830 feet; The Sentinel, 3,043 feet, with cascades of 3,000 feet fall, Washington Column, 1,875 feet; Dome and Royal Arches, 3,568 feet, down which descends a cataract of 1,000 feet; The Half Dome, 4,737 feet; Cloud's Rest, 6,150 feet; Glacier Point, 3,200; Sentinel Dome, 4,150; Eagle Point, 4,200, and many others of greater or less altitudes. The most noted waterfalls are the Yosemite, which first displays an unbroken descent 1,500 feet, then 600 feet of partly hidden cataracts, and a final leap of 400 feet—2,526 in all; Bridal Veil, 900 feet; Vernal Falls, 400, and Nevada Falls, 600 feet. There are many other points of interest, conspicuous among which are the Merced River, Mirror Lake, and romantic drives and climbs without number. There are a number of good hotels in the valley, and tourists are driven right up to their doors. The best time for visiting the falls is from the first of April until the end of July; but it is accessible until the snows of November close up its means of ingress and egress for several months.

The Big Trees. Thirty-five miles from Raymond is the Wawona Hotel (formerly Clark's) one of the most exquisite spots in the Sierra Nevada. There is an abundance of game near by, such as bear, deer (in great plenty), mountain quail, grouse and smaller game, while the adjacent streams abound in trout. It is from this hotel that tourists make their pilgrimage to the Mariposa Big Tree Grove, which is six miles, and is made in a carriage, and for which there is no extra charge for those holding through tickets to and from the Yosemite Valley. In this mighty grove there may be seen a large number of trees more than 300 feet in height, and varying from 50 to 93 feet in circumference, according to Professor Whitney's official measurement.

The Calaveras Grove, which was the first one discovered (by a hunter named A. T. Dowd, in 1852), has a magnificent lot of mammoth trees, also piercing the clouds at heights exceeding 300 feet, and measuring 80, 90 and 100 feet around at the ground. Most of these have marble slabs containing the names of distinguished soldiers, navigators, statesmen, poets, travelers and authors. The Calaveras Grove is 131 miles from San Francisco by rail, and 44 by stage—175 miles in all. The Mammoth Grove Hotel has lately been enlarged, and can now accommodate one hundred guests. There is a post-office, express and telegraph office at the hotel. It faces the grove, having the greater number of trees to the left, looking from the veranda, and the Two Sentinels immediately in the front, about two hundred yards to the eastward. The valley in which the hotel is situated contains of the Sequoia trees, ninety-three, not including those of from one to ten years' growth.

BIG TREES OF CALAVERAS

The *sequoia* is a representative of a family of trees, related to the cypresses, which has survived from a time more ancient than almost any other family of trees. Its nearest relative is in Japan. The name was given by the botanist, Asa Gray, in honor of Sequoyah, the Cherokee chieftain. Besides the *S. gigantea*, there is still another species, the *S. sempervirens*, which exists in forests along the seaward side of the Coast Range from San Francisco bay northward for over 100 miles. It is these forests which furnish the celebrated redwood lumber; and an illustrated article by Ernest Ingersoll, in Harper's Magazine for 1882, gives an admirable account of the lumbering operations by which these mighty trees are utilized, and of the interesting scenes in and about the region in which they grow. Many specimens of the redwood rival their big cousins near Yosemite in size, and the whole forest will average 250 feet in height, where full grown.

FROM SAN FRANCISCO TO THE GREAT NORTHWEST.

LONG reach of most interesting country lies between San Francisco and Portland, Oregon. Seven hundred and sixty-eight miles intervene between the two great cities, and it is our purpose to take the reader with us on this journey. There are two routes by rail; and, of course, the ocean highway is open to all who wish to go by steamer. The rail routes are east of the Sacramento River to Tehama, and west of the river to the same point, 125 miles from San Francisco, where the two lines form a junction. The route generally taken by tourists is that east of the river; and this is the route chosen for our journey. From San Francisco we return on the Overland route (by which we entered the city) as far as Junction, eighteen miles beyond Sacramento. Here we turn northward, leaving the main line behind us, and are fairly embarked on our journey to the Great Northwest.

Lincoln is a small manufacturing town, where great quantities of pottery and sewer pipe are made. (Population 600. Distance from San Francisco, 119 miles. Elevation, 167 feet.)

Passing through Sheridan, a village surrounded by grazing lands, we come to

Wheatland. Fitly named, it being in the centre of a fine wheat region. The town is well built, and has the usual complement of good business houses, churches, schools, etc. (Population 600. Distance from San Francisco, 130 miles. Elevation, 90 feet.)

The Yuba River. Leaving Wheatland we are soon crossing the bottom lands of what the latest maps call the Bear River, but which "old timers" know as the Yuba; a name which, it seems to us, should by all means be retained. The Yuba is here a vagrant stream, inclined to "spread itself" entirely too much for the convenience and comfort of the farmers; hence, it has been confined within great dykes, which extend as far as the eye can reach up and down the river. The road crosses the bottoms on trestle work.

MARYSVILLE.
Flourishing Commercial City.
County Seat of Yuba County.
Population, 6,000.
Distance from SanFrancisco,143 Miles
Elevation, 66 Feet.

This thriving place is the leading town of Northern California, the depot for the product of Yuba and Sutter Counties, and is situated at the head of navigation on Feather River and on the right bank of the Yuba. It has a population of 6,000. It is known throughout California as being the neatest built city in the State. Splendid business blocks; fine residences; magnificent gardens, where flowers bloom the year round; best of schools and academies; eight churches; large manufacturing interests; flour mills; finest woolen mill in the State; fruit cannery; iron foundry, etc. The city is lighted by gas and electricity. The water supply is considered the best in the State. The trade of Marysville to-day is greater than any town north of Sacra-

mento. It is the trade centre for a large country outside of Yuba County. It enjoys the trade of all Yuba and Sutter, and part of Butte, Colusa, Sierra, Placer and Nevada Counties. Two lines of railroad enter the town, and a third is now being pushed forward. Ten trains a day enter and depart. Two steamers and several barges ply on the river, carrying freight to and from San Francisco. It is one of the terminal points on the railroad. In climate, Marysville can not be be excelled. No extremes of heat and cold; but a pleasant, equable temperature, equal to, if not the superior of, the climate of Italy. Epidemic diseases of any kind never obtain a footing here; Marysville has been singularly free from such afflictions. With the fast increasing tide of immigration which is now turning to California, and with the new and varied industries which are now springing up here, as the producing power of the lands are becoming known, Marysville will, in a short space of time, no doubt, be one of the leading towns of California. Frosts are very rare, and when they do occur, very little damage to vegetation results, owing to the great

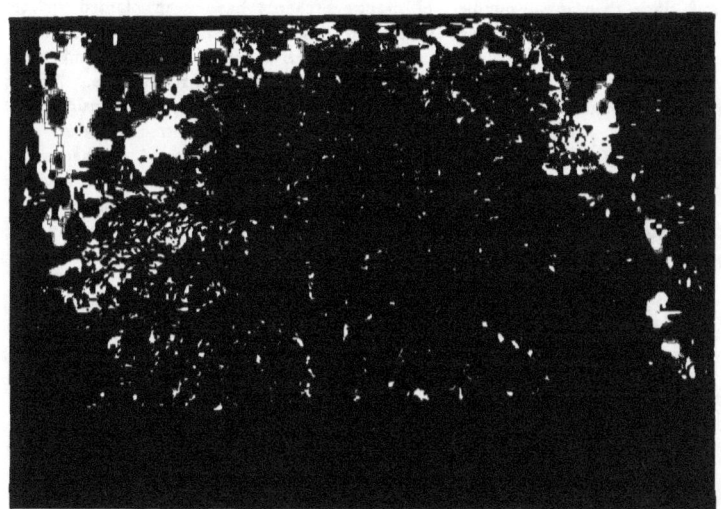

SIR JOSEPH HOOKER OAK, CHICO VECINO.
29 Feet in Circumference

dryness of the atmosphere. The same characteristics also make life very enjoyable, and render this section one of the healthiest in the State.

Oroville is situated on the Feather River, 28 miles from Marysville. It is the northern terminus of the Northern California Railroad, which runs from Marysville, 28 miles to the south. The town is well built, the business buildings being of brick, and the residences are almost universally neat and handsome, surrounded with lawns set with a wealth of flowers, palms and blooming orange trees. The church and school facilities are all that could be desired. One daily and two weekly papers are published. There is abundant water-power awaiting the establishment of manufactories, and a flouring mill and a large sash and door factory are now in operation. But the glory of Oroville is mainly in the region about it. The western part of Butte County, near the Sacramento River, is level, the eastern part includes the western slope of the Sierra Nevada Mountains, while the central portion consists of low foothills, gradually increasing in altitude, as the mountains are neared. This strip of sloping foothills, twenty miles in width, consists of a rich, gravelly soil, remarkably productive. The climate of this region, which is known as the "Thermal Belt," is of peculiar salubrity, being milder both in winter and summer than in the lower portion of the valley, and resembling that of the most favored countries about the Mediterranean Sea. The summer's heat is here tempered to an even mildness, and in the winter the formation of thin ice in the open air is of rare occurrence. Snow is a natural curiosity, and outdoor work is uninterrupted the year round. The average rainfall is about 22 inches. Experiment has shown that the conditions of climate and soil make this region the natural home of the orange, olive, lemon, fig and other semi-tropical fruits, while all the known deciduous fruits, including the hardy apple, flourish and yield in unsurpassed abundance. When, at the Northern California Citrus Fair, held in January, 1886, Butte County was awarded the first premium, the people of the county awoke to the fact that they lived in a fine orange-producing region, and since then great numbers of orange orchards have been planted. Each December since then a great citrus fair has been held at Oroville, and so marked has been their success, and so wonderful their revelations and their magnificence, that Oroville is rapidly becoming as noted as Riverside. While citrus fruits made up the most important features of these fairs, all the products of the county were also represented, and the Butte County Citrus Fairs are undoubtedly the greatest show of the fruits of the earth ever gathered together under one roof, including the fruits of nearly all climes, and all produced in one county. The country about Oroville is undoubtedly the greatest fruit-producing region in the State, offering great inducements to settlers, while it is equally wealthy in a great variety of other resources.

Returning to Marysville we resume our northward flight, the Sacramento Valley being on our left, while the Valley of the Rio de Los Plumas, or, as it is now popularly called, the Feather River Valley, is on our right. Following this course we pass through Live Oak, Gridley, Biggs, Nelson, Dunham, and arrive at

CHICO
An Ideal Residence City.
Population, 6,000.
Distance from San Francisco, 186 miles.
Elevation, 193 feet.

The largest town in Butte County, Chico, situated on Chico Creek, five miles from the Sacramento River, and on the line of the California & Oregon Railroad. Chico is the centre of the finest agricultural portion of the county — perhaps the finest in the State. The famous "Rancho Chico" property of Gen. John Bidwell adjoins the town on the north, the rich and varied fruits of which have attracted such marked attention at all fairs and expositions throughout the United States. Chico Creek is a clear and beautiful mountain stream, flowing sufficient water all the year to supply power for Gen. Bidwell's large flour mill, until its capacity was so enlarged as to require the supplemental aid of steam. Steamers run on the Sacramento River to Chico Landing and points above, carrying immense quantities of grain to the bay on barges. Chico is a beautiful city, and its population is principally American, agriculture and its adjunct employments being the chief elements of its life. But it has also tributary to it a fine mining region, up Butte Creek, and an immense lumber region to the east and north In this latter there are five or six large mills at work. A V-flume comes to the city from the mountains, in which the lumber is floated from the mills to the town so rapidly that a few years ago a beam of timber was sawn in the mill, thirty miles away, flumed to Chico, drawn through the town to the water-works building, fitted for its purpose and wrought into the building, all within the working hours of a single day. Chico has a regular town government, with police officers and an excellent fire department, which owns two steam fire engines. It has gas and water-works, and is supplied with electric light. There are two banks in flourishing condition. Seven churches, representing as many denominations, adorn the city, and two large and elegant public school buildings and two private academies are filled with children. The streets are wide, well kept and shaded. Very many private residences are large and handsome, and the homes of the people all indicate intelligence and comfort.

Chico Vecino. This is an attractive suburb of Chico, included within the boundaries of the well known Rancho Chico. There are one thousand acres in the town site, the plat of which has been laid off in five-acre tracts. Here there will soon be one of those delightful fruit-raising colonies for which California is becoming famous. From Chico to Tehama we roll along through a fine fruit and agricultural country, passing the stations of Vina, Nord, Anita, Cana, Soto and Sesma.

Tehama is the junction of the Willows Branch of the Southern Pacific Railroad with the main line. It is situated in a good wheat-growing country on the west bank of the Sacramento River, and here irrigation is not found necessary for the production of crops. Stock-raising and lumbering are large tributary industries. (Population, 329. Distance from San Francisco, 213 miles. Elevation, 222 feet.)

Seven miles beyond Tehama we pass through Rawson, and five miles farther on reach **Red Bluff,** the county-seat of Tehama County which is one of the most thriving towns of the State. It is a growing town in one of the richest sections, and it has an elevated and sightly location. Its streets are wide and well graded, lighted by electricity; and there is no place in the United States better drained The Sacramento River here is a clear, rapid stream, lined with beautiful trees and vines. On all the three other sides there are ravines or valleys through which streams run, which give the

ON THE RIO CHICO.

perfection of drainage. Its public and business buildings are fine architectural structures; and its private residences are nowhere excelled for taste, elegance, and the beauty and the wealth of their floral surroundings. The streets are lined with popular, elm, white maple, locust, acacia and pepper trees, which will soon make a veritable forest city. There are also many fine residences. Tehama County is the great grain-growing county of the State; 8,000,000 bushels of wheat and 2,500,000 bushels of barley have been harvested in one season from its fertile lands Tehama has about 400,000 sheep, which produce 2,500,000 pounds of wool annually. The numbers of cattle, horses, mules and swine are large. In this county the celebrated Vina Ranch is located, embracing 56,000 acres, a

NAPA SODA SPRINGS, CAL.

princely property, which, through the unexampled generosity of Senator and Mrs. Stanford, has become the heritage of the children and of the coming generations of the Pacific Coast. (Population, 3,500. Distance from San Francisco, 225 miles. Elevation, 307 feet.)

The grade is now steadily upward as we press onward in our journey. From Red Bluff to Sissons, a distance of 113 miles, we make an ascent of 3,245 feet.

TWIN FALLS.

Through a broken country, and crossing a number of rapidly flowing creeks, we pass through Hooker and Cottonwood (small stations) and arrive at

Anderson. Which is a beautiful and very lively town of 1,500 inhabitants, on the line of the recently completed California & Oregon Railroad. It lies a mile and a half from the Sacramento River, 8 miles south of Redding, and 222 miles north of Sacramento. The town is attractively laid out, with wide, well shaded streets, lined with cosy and beautiful homes. The leading hotel in the place is a fine one, costing $20,000. There are fine schools; the usual churches: a fine roller flouring mill; good, substantial brick business buildings; water works, furnishing an abundant supply of pure water from the mountains; and a live weekly paper, besides many other evidences of enterprise and progress. The semi-tropical climate of the Sacramento Valley generally prevails in the region about Anderson, which is noted for its healthfulness. The summers are rather warm, though dry, and the mercury rarely reaches 105 degrees, 85 degrees being about the average. The winter, or rainy season, is delightful, and resembles April or May in the Eastern States. (Population, 750. Distance from San Francisco, 249 miles. Elevation, 432 feet.)

Redding. No town of Northern California has a more promising future, and exhibits at the present time more enterprise, activity and rapidity of growth than Redding, in the southwestern part of Shasta County, of which it is the county-seat. It is at the upper end of the great Sacramento Valley, 169 miles north of Sacramento, and is built on a plateau on the bank of the Sacramento River, here a clear mountain stream which sweeps around the town to the east and south. No town in the State has a more charming and picturesque location. The brief history of Redding is one of rapid progress, and never has it been more marked than now. Its population has increased from 500 in 1883 to over 2,0co at the present time, and with the rapid development of the county, which will follow the recent completion of the first railroad through this region, and the vast territory that must remain tributary to Redding, extending in some directions a hundred and fifty miles, a rapid and continued growth is assured. The city has water and gas works, a great variety of manufactories, many important buildings, a fine court house and jail in process of construction, two newspapers, good schools and several churches. The river here affords fine water-power and the lumber interests of the country tributary to Redding are immense. The future of this lively place depends largely on the development of the country about it; and with the great variety of soil, climate and products, the thousands of acres of cheap, unoccupied lands that only await intelligent cultivation to yield great profits, and with the other almost inexhaustible resources which the county possesses, there can be no question on this point. During the past year the county has made rapid strides many settlers have invested, building has amounted almost to a boom, new industries started, and thousands of acres of orchards and vineyards have been planted. No part of California offers such inducements to the farmer, the laboring man, the capitalist, or the home-seeker, as Shasta County. There is a delightful semi-tropical climate in the valleys and plateaus of the south, and a gradual change is noted as higher altitudes are reached, that of the mountains resembling the New England States. The climate of the southern portion of the county is indicated by the fact that orange trees flourish and bear abundantly. The county is noted for the number and beauty of its clear, sparkling streams, which burst from the mountains through wild, picturesque cañons, and flow onward through small fertile valleys of great beauty. In these mountain streams the finest trout-fishing in the State is found. (Population, 2,500. Distance from San Francisco, 260 miles. Elevation, 551 feet.)

Wild Scenery. After leaving Redding our course is directly toward the Shasta Range of mountains, and the scenery grows in grandeur as we advance. Within a distance of 80 miles we cross the Sacramento River eighteen times, and pass through just an even dozen of tunnels. Grander and grander grows the scene as we advance. The roll of stations as given in the railroad time tables gives no idea of the beauty which surrounds these villages, but as a matter of record we will name them as follows: Middle Creek, Copley, Kennet, Morley, Elmore,

Smithson, Delta, Slatons, Gibson, Chromite, Sims, Castle Crag, Lower Soda Springs, Chestnut, Dunsmuir, Upper Soda Springs, Cantara, Mott, McCloud and Sisson. The Soda Springs mentioned above are of interest to the tourist and health seeker, being medicinal in their qualities, and having good hotel accommodations. The Lower Spring is two miles below Dunsmuir, and the Upper Spring one mile above. Beyond Upper Soda Springs we pass through a tremendous gorge, whose beetling crags tower above our heads, and before McCloud is reached we circle the Big Bend of the Sacramento River, traveling over five miles to gain an advance of half a mile; but it must be considered that we have also gained an additional elevation, nearly six hundred feet.

> **SHASTA.**
> The Monarch of the Range.
> Altitude: 14,440 feet above the Sea.
> Local Elevation: 10,885 feet.

As we near Sisson, Mount Shasta, of which we have obtained brief glimpses through the pines, bursts into full view in all its sublime magnificence. This noted snow capped peak, towers to the height of 14,440 feet. It is an extinct volcano, and its snows and glaciers feed hundreds of streams which thread the wild region in every direction. Sisson is a regular meal station, situated in Strawberry Valley, one of the most beautiful vales of California. It is a new but rapidly growing town, with stores, hotels, a weekly paper, a fine depot, and a roundhouse and repair shops. Here is obtained the finest view of Shasta, and it is the only convenient point from which the ascent can be made. But few parties succeed in reaching the summit, and the attempt is only made in midsummer and then with trusty guides. The feat is perilous and exciting, but the view is grand beyond description. The region about Sisson is a paradise for the sportsman and the lover of nature. Grizzly, black and cinnamon bears abound; elk, deer and mountain sheep are plenty, as well as a great variety of smaller game. The mountain streams teem with trout, and often the sport loses its zest through the very abundance of the beauties. The McCloud and the Pitt Rivers are the most noted streams, though others are equally attractive. The McCloud runs through the most uninhabited and unexplored region on the coast. No region in the State is so delightful for camping, and hundreds of parties go there every year. At Sisson, camping and hunting parties can be provided with complete outfits at moderate cost. The pioneer of Strawberry Valley is J. H. Sisson, from whom the town derives its name. He knows the whole country thoroughly, and has taken many parties to the summit of Shasta. His "tavern," with its quaintness, its great fireplace and its hospitable welcome, is in perfect accord with the spirit of the tourist and the surroundings, and enjoys a wide reputation. As has been said, Sisson is situated at the foot of Mount Shasta, and is noted for its magnificent scenic attractions. From Redding northward the California & Oregon road is the scenic route of California; and at Sisson, at the base of Mount Shasta, 80 miles north of Redding, the acme of interest is reached. While there are many places in California replete with beauty and grandeur, there are none which, for infinite variety of scenery, wildness and abundance of every thing to delight the sportsman, artist and tourist, can compare with the region about Sisson. (Population, 250. Distance from San Francisco, 338 miles. Elevation, 3,555 feet.)

Muir's Peak. After leaving Sisson we circle the base of Muir's Peak, locally known as "Black Butte," which rises to a perpendicular height of 3,000 feet above our heads. It is black, bare and desolate,—an extinct volcano, with

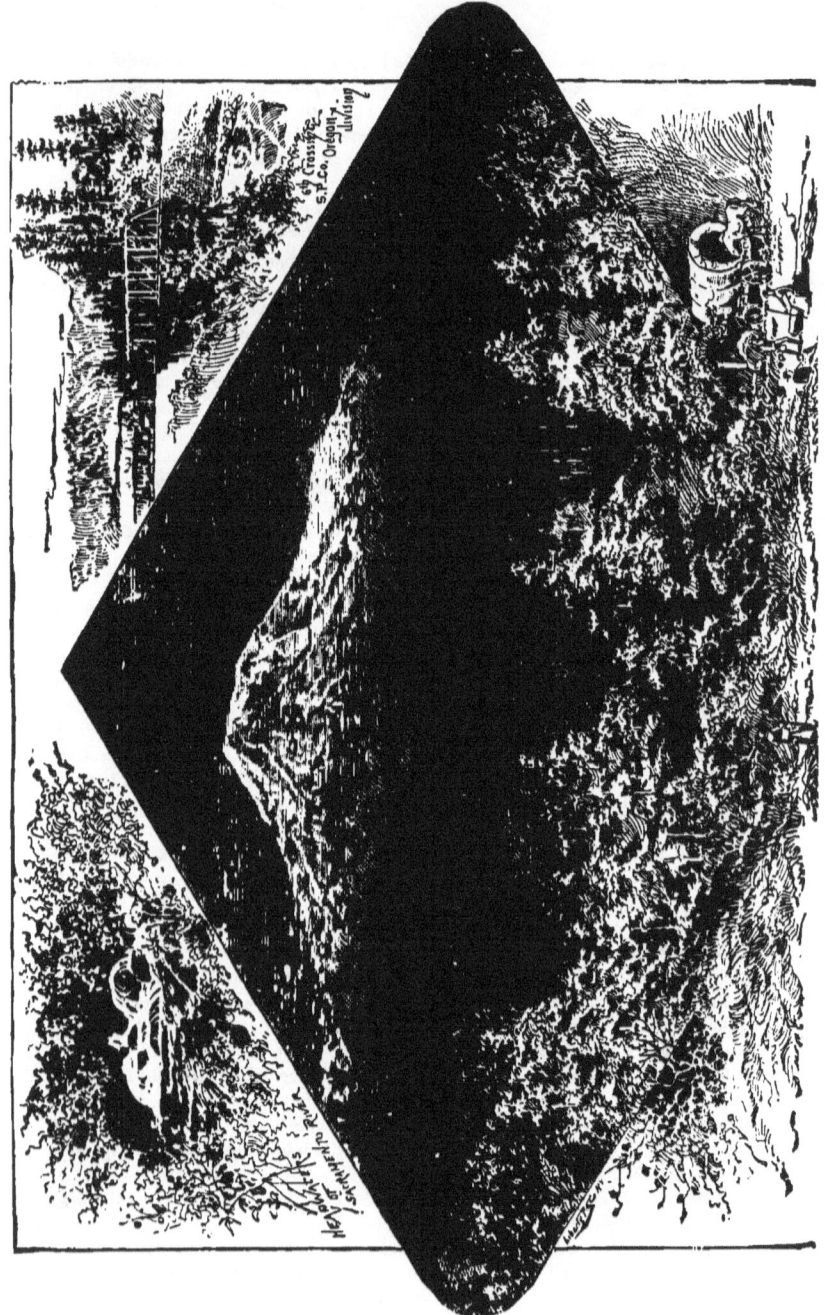

half a dozen craters in plain view. We are now among the mountains, and pass in succession the stations of Igerna, Edgewood, Gazelle, Grenada, Montague, Snowden, Ager, Hornbrook, Zuleka and Coles.

The State Line. Two miles beyond Coles station we cross the State line, and, entering Oregon, begin the ascent of the Siskiyou Mountains. This ascent is a wonder of railway engineering. The statistical facts concerning this achievement may be condensed as follows:

Elevation at State Line	2,859	feet
" Coles Springs	3,775	"
" Tunnel No. 13	3,108	"
" Tunnel No. 15	3,710	"
" Tunnel No. 16	2,977	"
Length of Tunnel No. 13	4,160	"

The mathematician has the advantage here. He can tell exactly the facts concerning this great work; but the descriptive writer strives in vain to convey to the reader the beauty and grandeur of the scene. The southern slope of the range is denuded of trees, while the northern side is covered with a dense growth of pine.

Siskiyou Station. This is the summit of the range, and the highest point on the entire line, being 4,135 feet above the level of the sea. The mountain view from this coign of vantage is indescribably magnificent. To the east is the Cascade Range, extending to the north for full four hundred miles; to the northeast is Mount Pitt while still farther on are Mounts Scott, Threlson and Diamond Peak,—monarchs of the Cascades. To the west are the peaks of the Siskiyou and Coast Ranges; to the south are the Two Sisters, Mount Lassen, and above all imperial Shasta rears his head. Lakes, rivers and valleys lie spread out before us like a map; and, in a word, for variety, grandeur, beauty and extent, this view has no equal on the continent.

Ashland. At the foot of the Siskiyou Range, on the eastern slope, is situated this beautiful little town, in a delightful valley. The town was established in 1850; and in 1887, on December 17, Mr. Charles Crocker, of San Francisco, drove the last spike which completed the railroad connection between California and Oregon. The town of Ashland has entered upon a season of great prosperity, being the seat of the State Normal School, and having the White Sulphur Springs within near proximity. It is a large shipping point for wheat, and also for fruit. (Population, 2 000. Distance from San Francisco, 431 miles. Elevation, 1,891 feet.)

Rolling along through the valley we pass Phœnix and Medford, prosperous towns of moderate size.

Jacksonville is the county-seat of Jackson County, and is connected with Medford, four miles distant, by stage. (Population, 1,200. Distance from San Francisco, 450 miles. Elevation, 1,399 feet.)

Rogue River Valley. We are now in the Rogue River Valley, and are following the stream in its downward course. The valley averages about three miles in width, with high hills on each side, covered with a strong growth of grass and in places heavily timbered. The products of this valley are berries, nuts and fruit. Fishing and hunting can be found here of the best quality. The stations which follow Medford, are: Gold Hill, Grant's Pass, Merlin, Aeta, Almaden, Glendale, Riddles, Myrtle Creek, Oak Grove, Dillard and Greens

"THE OLD CABIN, ON THE COLUMBIA."

SCENIC ATTRACTIONS.

A Panorama of the Grand and Beautiful in Nature.

For a stretch of over one hundred and fifty miles from Grant's Pass, the country presents a wonderful panorama of grand and beautiful scenery. Mountains are all around us. To the right the Cascade Range, to the left the Coast Range. Gorges before us!—cañons behind us! Little valleys of entrancing loveliness are crossed; sparkling streams abound, forests of oaks and pines, of hemlocks and madrones are threaded; in a word, the variety is infinite, the beauty indescribable.

Roseburg is the county-seat of Douglas County. Through the town flow the Umpqua River and Deer Creek, which furnish water-power and a plentiful supply of pure water for domestic purposes. Agriculture, horticulture and pastoral industries are tributary. (Population, 1,500. Distance from San Francisco, 574 miles. Elevation, 487 feet.)

The Valley of the Umpqua. This valley, situated between the Coast Range of mountains and the Calapooias, is exceedingly fertile, being especially adapted to agriculture and the growing of fruit. The valley ranks third in size among those of Oregon, those of the Willamette and Umatilla being greater in area. A historical interest attaches itself to the Umpqua Valley, for in its quiet confines lie the remains of the brave soldier and public-spirited citizen, General Joseph Lane. His grave is in a little churchyard, a mile from Roseburg. After leaving Roseburg, the stations occur in the following order: Wilbur, Oakland, Rice Hill, Youcalla, Drains and Comstocks.

Divide is on the water-shed between the waters of the Umpqua and Williamette Rivers. Latham, Cottage Grove, Walkers, Creswell, Goshen and Springfield are the succeeding stations.

Eugene is the county-seat of Lane County, situated on the right bank of the Willamette River, and is a thriving, prosperous town. Here has been established the University of Oregon, which is one of the leading educational institutions of the State. The Willamette is navigable from Portland to this point for steamers of light draught; but freight traffic is now carried mainly by the railroad. This is a fine agricultural and fruit country, and shipments of these products from Eugene are large. (Population, 2,200. Distance from San Francisco, 649 miles. Elevation, 455 feet.)

Beyond Eugene are Irving, Junction City, Harrisburg, Muddy, Halsey, Shedds, Tangent and Albany Junction.

Albany, the county-seat of Linn County, is an enterprising, growing town. For a country which eastern people consider so "new," this town has great "antiquity," having been established in 1848. Here is located the Albany College and other schools of excellent quality. The town has good business and private buildings, water works,—in fact all of the modern improvements. (Population, 3,000. Distance from San Francisco, 692 miles. Elevation, 240 feet.)

Millers, Jefferson, Marion and Turner are the stations passed after leaving Albany before Salem is reached.

Salem is the State capital and the county-seat of Marion County. It is situated on the left bank of the Willamette River, which furnishes unlimited water-power. Here are located the State institutions, including the Insane Asylum, the School for the Deaf and Dumb and Blind, the Penitentiary and the Indian Training School. Steamers ply regularly between Portland and Salem, and the amount of lumber shipped annually exceeds three million feet. The Capitol Building occupies an entire block, and may be seen from the car windows, to our left, after leaving the depot. It need not be said that Salem is a well-built, prosperous city, for the fact that it is the State capital makes such a statement superfluous. (Population, 6,000. Distance from San Francisco, 720 miles. Elevation, 190 feet.)

After we have left Salem we pass the State Fair grounds, two miles from the city, and four miles farther on Chemawa is reached, which is the immediate site of the Indian Training School. Beyond are the stations of Brooks, Woodburn, Hubbard, Aurora, Barlow, Canby, New Era and Canema.

Oregon City is the county-seat of Clackamas County, and is noted for its magnificent water-power, being located at the great falls of the Willamette River. Here were constructed the canal and lock system which make the Willamette navigable beyond the falls. This system cost half a million dollars. Oregon City is a thriving town boasting all the modern improvements, and doing a large business. (Population, 1,400. Distance from San Francisco, 575 miles. Elevation, 95 feet.)

Beyond Oregon City we pass through the following stations: Clackamas, Milwaukee, Wellsburg, Machine Shops and East Portland. These are really suburbs of Portland, as the distance between Oregon City and Portland is only 15 miles.

PORTLAND.
The Metropolis of the Pacific Northwest.
A City of Magnificent Achievements and High Hopes.

This metropolitan city, with its population of fifty thousand souls, sits on the west bank of the Willamette River, twelve miles from its confluence with the Columbia, and one hundred and fifteen miles, by river, from the Pacific Ocean. The first settlers came here in 1843, and in 1851 the settlement was incorporated as a city. It is now the metropolis of the Pacific Northwest, and the third richest city in the world, in proportion of the wealth to per capita of population. On the east side of the Willamette, directly opposite Portland, is the city of East Portland, and on the same side, to the northward, around the bend of the river, the city of Albina, both of which contain a population of about ten thousand souls, and are connected with Portland by two bridges. Numerous ferry boats also ply on the river between Portland and her trans-Willamette suburbs. Thus, there are clustered here under three corporate names, a community of sixty thousand people, whose business intermingles, and who are actively engaged in its diversified industries.

The favorable position which Portland occupies for an important commercial city, can be best understood by gaining a knowledge of its location, relative to a large area of very rich country. The Willamette Valley, at the foot of which Portland is situated, contains four million acres of land, and its products are abundant to furnish sustenance for over a million people. Most of this territory is now under cultivation. Wheat has been the chief crop raised, but other cereals, root crops and fruits are now occupying the attention of the farmers, and on the slopes of the mountains that border the valley, stock raising and dairying are found to be profitable industries. The finest flavored fruits in the world are raised here—apples, pears, prunes, peaches, plums, small fruits, melons, etc. In fact, all the products of the temperate zone can be successfully grown in the Willamette Valley. The surplus product of this fertile valley, of course, flows through Portland, to which port it is transported by boats which ply on the Willamette, and railroads which penetrate the country on each side of the river.

The Columbia River, before piercing the Cascade Mountains, flows through and drains a tract of country more than four times as large as the State of New York, and with a soil of wonderful productiveness. The improvement of that vast region is scarcely begun, yet the product has already grown beyond the facilities for moving it, though they are great, and beyond all expectations. But the transportation facilities are increasing rapidly, and that trouble will not last. Anything that can be grown on fertile soil in a mild climate is produced in this basin in abundance, and from Idaho, Washington Territory and Oregon, a constant stream flows to Portland.

The mines of Oregon, including those of gold, silver, iron, copper, etc., and the vast mineral output of Montana, Idaho and Washington contribute an important amount to the business of this commercial metropolis. The timber product is by no means inconsiderable, large quantities of lumber being annually turned out. The most extensive salmon fishing in the world, and the general piscatorial industry of the Columbia and Willamette Rivers, have their main springs of capital in Portland. Situated as she is, at the gateway to the regions mentioned, the resources of which are practically illimitable and easily transported on the rivers that drain them, being accessible to ocean craft, and having a demand for trade from across the sea, being at a point of interchange of foreign and domestic traffic, having a situation favorable for utilizing these various agencies for promoting growth, Portland certainly possesses advantages of location equaled by few cities in the world.

PORTLAND FROM EAST BANK OF THE WILLAMETTE.

There are five lines of railroad centering in Portland. The Northern Pacific runs north to Tacoma, thence east to St. Paul. It also connects, at Wallula Junction, with the O. R. & N., making a shorter route from Portland to the East. The Oregon Railway & Navigation Company has a line passing up the Columbia River to Wallula Junction, and branching out into various feeders, built and in

CAPE HORN, COLUMBIA RIVER.

LOWER CAPE HORN, COLUMBIA RIVER.

process of construction, ramifying the south-central portion of the great Inland Empire. The lease of the O. R. & N. to the Union Pacific has been consummated, and this now gives a direct trans-continental line under one management between Portland and Omaha. The Southern Pacific Company has leased the Oregon & California Railway, which has been completed, and this, besides affording a rail route between Portland and San Francisco, gives a through line, under one management, from Portland, via New Orleans, to New York. This road runs southward through the Willamette Valley. Another line of the Oregon & California starts from Portland, and, running up the west side of the river, forms a valuable feeder, penetrating the heart of the garden of Oregon. This line connects, at Corvallis, with the Oregon Pacific, extending westward to Yaquina Bay, and will soon reach a rich but as yet undeveloped region in Eastern Oregon. Then the Portland & Willamette Valley narrow gauge affords another outlet for the valley through Portland. Thus, this city is made a terminus for three trans-continental railway systems and has all the advantages of five local roads, besides the water transportation on the Willamette and Columbia Rivers and the Pacific Ocean. The Canadian Pacific is also competing for Portland business, running a steamer between here and Vancouver, B. C., to connect with its China line of steamers, and bidding eagerly for freight and passenger business between Portland and the Eastern States. The Northern Pacific Terminal Company has erected shops in Albina, at a cost of over $500,000, with a capacity for the employment of a thousand men. The company owns nearly eight thousand feet of water front. Besides the shops, there are large grain warehouses, coal bunkers, and a dry dock, owned by the Oregon Railway & Navigation Company.

On the Portland side of the river, about thirty acres of land have been purchased for a site for union passenger and freight buildings, and for a freight yard. The completion of the bridge over the Willamette, which the Oregon Railway & Navigation Company has constructed, enables the improvements contemplated for the Portland yard and buildings to be carried out. This bridge is a steel structure, consisting of a draw span of three hundred and forty feet and a fixed span of three hundred and twenty feet. It is a through bridge, with carriage way and footwalks above the railroad tracks, and connects Third Street, Portland, and Holladay Avenue, East Portland.

Modern Improvements. The streets of Portland are lighted by six hundred incandescent and twenty-four arc electric lights. The city owns its water works system, on which $500,000 have been expended, and improvements necessitating the expenditure of $125,000 are contemplated. In order to purchase the water works plant from the private corporation which owned it, the city issued five per cent. bonds to the amount of $500,000, which were readily sold at an average price of $1.08, showing the confidence in the city's financial condition. The city has thirty-two miles of water mains, and the pumping capacity of the works is fifteen million gallons per day. The supply is obtained from the Willamette River, about five miles up the stream. The average daily consumption is five million gallons.

The Portland Paid Fire Department is an efficient organization, operating under the City Board of Fire Commissioners. The official report for 1887 shows the value of real property and apparatus held in trust by the organization to be $171,350.21. The total running expenses for the year were $58,927.69. There are thirty-three electric fire alarm boxes, and the system is in excellent working order. The numerical strength of the Department is ninety men. There are twenty horses, seven engines, with their necessary adjuncts, and seven thousand feet of rubber hose. The losses by fire during the year amounted to $84,173.72, for which $80,311.62 were paid in insurance. A fireman's mutual relief association is in operation in connection with the Fire Department.

Manufacturing. The manufacturing advantages of Portland and vicinity are not utilized to an extent at all commensurate with their importance. There is abundant raw material in Oregon, cheap and reliable water power, and generally favorable conditions for the growth of varied manufacturing enterprises. The comparatively recent discovery of the resources of the region must account for the small amount of manufacturing that is done where circumstances are so favorable. People from the East, accustomed to the closer and fuller development of their resources, and alive to the advantages of manufacturing as near the source of supply as possible, are surprised at the neglected opportunities which they observe on the Pacific Slope, and particularly in and about the commercial centre of a region incalculably rich in the elements that promote manufacturing prosperity. Still, that branch of industry is well established, and is constantly increasing in volume and importance. (Population, 50,000. Distance from San Francisco, 772 miles.)

Picturesque Surroundings. Aside from the advantages of its relative location, Portland has a very admirable site for a beautiful city. From the docks at the river's side, the land gradually ascends to the west and southwest, finally breaking in elevated and picturesque hills, upon which the residence portion of the city is already encroaching. These hills form an important feature in the topography of the city. The lower and more level part of the town is occupied by business houses and manufactories. The heights are visible from almost any point. They are ascended by means of roadways winding along the hillsides, affording magnificent views

FORESTS ON THE COLUMBIA.

as the prospect unfolds. From the summit of Robinson's Hill, on a clear day, the sight is most grand and inspiring. Within a radius of hundred a miles, which the eye sweeps from this elevated outlook, north, east and southeast, five perpetually snow-clad mountain peaks are visible. The most prominent of these is Mount Hood, which rests upon the long, bluish bank of the Cascade Mountains, and rears its lofty summit to the sky. Its covering of snow and

ROOSTER ROCK, COLUMBIA RIVER.

glaciers sparkles in the sunlight, and when suffused with the soft glow of the setting sun, reflects the most delicate tints of purple, crimson and gold, giving it a majestic splendor inspiring to the beholder. To the south is Mount Jefferson, and to the north Mounts Adams, St. Helens and Rainier, the latter the loftiest peak of the Cascade Mountain Range, all of them capped with snow and ice, and relieving a landscape of charming beauty. Breaking through the ridge of the Cascades, the great "River of the West," the Columbia, pours its mighty tide toward the sea. The Willamette threads the broad valley to the south like a ribbon, its course being visible for many miles and finally being lost among the farms and villages that dot its banks. For further description of this city, the reader is referred to the Addenda.

TACOMA.

A City whose Fame has become International.

"The City of Destiny."

Tacoma's commanding position among the cities of Washington Territory has been earned step by step by a struggle in which the odds were against her. The general apprehension, justified probably by the history of many cities and towns, that in the West all one need to do is to stake off a few lots, build a cabin or two, select a name, and a city will grow up much after the fashion of vegetables in a garden, is in nowise true of Tacoma. When Tacoma was established, other towns on Puget Sound had existed for many

DIP-NET FISHING AT THE DALLES OF THE COLUMBIA   By permission of the West Shore Magazine.

years, and naturally they did not extend any encouragement to a new town. Instead of receiving from the beginning, as in the case of many cities of the West, the exclusive support and encouragement of an extensive business district, Tacoma found the older towns already in possession, and ready to contest every step taken by the new claimant for public favor and support. Figuratively speaking, Tacoma's first breath of life was a battle-cry, and although the cry was not at first very loud, it was firm, full of confidence and pluck. The town did not remain long in its swaddling clothes. Its voice gained in strength. At first Puget Sound only heard it. Then it reached the ears of everybody in Washington Territory, and they were pleased with it. The Pacific Northwest then realized that there was a new voice in the business world and stopped to listen, and soon the entire Pacific Coast was talking about it. Then the great and populous East heard Tacoma's voice, and when it said "Come," thousands responded. Then England came thousands of miles by sea, in great ships, to learn more about Tacoma, a city whose fame had crossed the Atlantic. China and Japan sent tea-ships at this infant's demand, and even far-off Australia heard it, and was so pleased that the ocean pathway between Tacoma and that continent is marked by an ever-increasing fleet of ships going and coming. Tacoma helps to feed the world; helps to build the world's houses; and yet its voice is stronger than ever and is being used more than ever. The thousands of people who listened and responded to Tacoma's invitation were not disappointed. And Tacoma grew and flourished until its present commanding position was reached.

From a town of only a few hundred people, Tacoma now has a population estimated at 15,000. Its property has increased to a dozen times its value eight years ago. Its business relations extend to all parts of the civilized world, a fact which is true of no other city in Washington Territory. In railroads, shipping, manufactories, and business generally, Tacoma's prosperity has been very great; so great, indeed, that whereas it a few years ago was only a small and relatively unimportant village, it is now a city, possessing all the characteristics and conveniences of a city.

Tacoma was originally planned on a large scale, and the expectations of the founders of the city, however sanguine they may have been, have doubtless been more than realized at this time. Probably no one expected Tacoma to grow so rapidly, to earn so speedily such extraordinary trade relations with the markets of the world. The streets are wide and laid off with special relation to convenience and beauty. Pacific and Tacoma Avenues are without superiors for beauty and length in the Northwest. These and other public highways are well graded, and sidewalks are constructed of a substantial character.

The location of the Methodist University in Tacoma has given the city a notable addition to its already large number of educational institutions. The Tacoma people subscribed a bonus of $75,000 to this great institution. The Annie Wright Seminary, the Washington College, and the numerous public schools, speak more than words can tell of the public spirit manifested by Tacoma people, of their ability to meet every demand of a liberal and progressive population and of the existence of a breadth of public sentiment which proves the stable character of the city's progress. The need of a street railway has for a long time been been a pressing one, and the result is that now a well-constructed and equipped road, about three and a half miles long, is in operation. It extends from the docks to the centre of the city. Of the many church buildings, some possess architectural beauties equal to those to be seen anywhere. Private residences of handsome architecture may be seen in all parts of the city. The hotels number twenty, and yet they are not

PACIFIC AVENUE, TACOMA. 1877.

PACIFIC AVENUE, TACOMA. 1888.

sufficient to accommodate the multitude of people who daily arrive in this flourishing city. Tacoma has no indebtedness. This tells a volume itself.

The Northern Pacific Railroad Company has erected a magnificent brick building for the offices of the company.

These features of Tacoma are worthy of special attention as evidencing the solid character of the city's progress. They rebut every idea that Tacoma's growth and the expansion of her industries are "mushroomy" in character. The city itself is the best commentary on the character of its resources. (Population, 15,000. Distance from San Francisco, 917 miles.)

The Climate of Puget Sound. The following extract from a recent compilation so accurately sets forth the characteristics of this climate, that to employ other words would add nothing to the facts contained in it:

The climate of the Puget Sound country is wholly unlike anything experienced on the Atlantic Slope, or in the Mississippi Valley; or, indeed, anywhere on the American continent except in the Pacific Northwest. The summers are cool and the winters singularly mild. A temperature of 80° in midsummer is very rare, and not often in winter does the mercury go much below the freezing point. The following is the meteorological table for 1885, which is about an average year, compiled from observations taken daily at 7 a. m., 2 p. m. and 9 p. m. A minute's study of it will show how remarkably free from trying extremes the climate is.

METEOROLOGICAL TABLE FOR 1885.

	Lowest.	Highest.	Mean.	Monthly Rainfall in inches.
January	30°	62°	37.9°	4.20
February	31	59	44.5	4.16
March	32	68	48.0	1.01
April	35	75	50.8	0.47
May	43	80	60.5	2.89
June	47	76	57.0	0.49
July	51	86	66.4	0.26
August	52	84	64.5	---
September	46	74	58.8	2.44
October	39	64	51.4	2.47
November	34	60	45.5	8.22
December	28	60	41.4	6.14
Total rainfall in 1885				32.74

As suggested, if the above extract is carefully studied it will tell more than many words of explanation.

Trade with South America and Mexico. The condition upon which trade relations will be established with South American and Mexican Pacific Coast points are of such a promising character that it will not be long until a most valuable commerce will be carried on. The peculiar conditions which justify the hope of establishing very extensive relations with that country are found in the products of the countries. The purposes of this article will not admit of a minute examination of these conditions, but any one who will examine the subject will find that the products of Washington Territory supply what the South American countries referred to do not have, and those countries produce that which will find a ready market in the Northwest. Hard woods, tropical fruits, valuable ores and minerals on the one hand, with soft woods, iron, grain, fish and many other of the resources of the Northwest—these, any one can easily see, furnish all the conditions upon which most extensive commercial relations may be established. The

CICERO PLACE, ON GREEN LAKE, NEAR SEATTLE, W. T.

relations will be those of exchange of products. Such conditions are especially promising, as they will afford cargoes both going and coming.

Tacoma's commercial relations with the Pacific Coast are now so well known that it is almost unnecessary to make reference to them at all, except to make this array of evidence complete. Reference to the record of Tacoma's shipping, as set forth already in this article, will show how extensive are Tacoma's relations with San Francisco and other coast points.

The thoughtful man will reason that if Tacoma enjoys such extraordinary advantages now, what will the future bring? He will then understand the peculiar significance of the poetical phrase, "The City of Destiny."

A Magnificent Harbor. The general measure of Tacoma's appreciation of this most remarkable body of water would be expressed in miles rather than particular instances. To say that there are saw mills at particular points, coal bunkers at others, wheat warehouses near by, magnificent docks elsewhere, various harbor improvements and railroads, would certainly be very suggestive of what Tacoma has accomplished in a few years. But to say that these improvements extend along the water front for a distance of about six miles, gives a larger idea of their extent.

These features of Tacoma's enterprise and prosperity have a special meaning. They are not constructed simply as a matter of ornament. Business men do not do things that way. Business methods are not fancy in their character. These improvements indicate that demands exist and are being supplied. And Tacoma is doing the supplying.

Terminal and Shipping Facilities. The fact that the Northern Pacific Railroad has made Tacoma its terminal point, is of itself enough to satisfy anyone, without further explanation, that the terminal and shipping facilities would be commensurate with the importance of a great trans-continental railroad company's interests.

The immense docks at which railroad and ocean traffic unite, are so large and involve so many distinct features, that it would be difficult to impart to anyone not familiar with such improvements an adequate idea of their extent and importance.

It is not an uncommon sight to see lying along these immense docks, only a few feet away from the railroad tracks, an ocean sailing-vessel, several ocean steamships Alaska steamers, besides a host of smaller craft. This will suggest the character and extent of these docks. The Northern Pacific Company has immense warehouses erected on these docks, and all the conveniences incident to the prompt, careful and expeditious handling of freights. It is often a difficult matter for local craft to secure dock accommodations, so crowded with steamers and sailing-vessels do the docks become.

The conveniences are such that the handling of immense cargoes is accomplished with an ease and dispatch scarcely conceivable. The ships laden with tea are drawn up within a few feet of the great warehouses, alongside of which are the railroad switches. The San Francisco steamers also discharge their freight into these warehouses Extensive additions have been made to these docks within the past year, to accommodate the ever-increasing demand for room, and more extensions are in contemplation.

Trade with the Middle West. The trade with the Middle West and in the Far East is made up of tea and lumber and shingles. In lumber and shingles most promising trade relations have been established with the sections referred to, and the trade in these products is constantly increasing in volume. The excellence and durability of the cedar shingles manufactured in Tacoma and

MEGDENHOUR BAY AND EDGEWATER POINT, NEAR SEATTLE, W. T.

vicinity make them superior to any manufactured elsewhere, and large quantities are now being shipped East. The qualities of Puget Sound lumber has made it famous all over the world. Tacoma being the terminal point of that great transcontinental artery of commerce the Northern Pacific Railroad, naturally enjoys the results of such special advantages. It does not require elaborate reasoning to convince any man that the same conditions which gave rise to such trade will increase its volume rapidly the longer the relations exist.

Tea Trade with the Orient. It is less than a year since the first tea ship arrived in Tacoma from Yokahama. It was only a few months after the completion of the Northern Pacific Railroad Company's Cascade branch. This shows how quickly Oriental and American merchants realized the advantages attending the shipping of tea to Tacoma. The great gain in time and reduction in expense were the considerations which have brought to Tacoma such an important branch of San Francisco trade. Two ships at this writing are on the ocean with tea cargoes from the Orient to Tacoma. Is there need to expand on the significance of these relations? Is there need to repeat the fact that trade brings trade? Tea and lumber will not always be the only articles of commerce between the Orient and Tacoma. This is only the beginning, and it does not require much imagination to picture in the near future a constant stream of vessels, both steam and sail, between Tacoma and the various commercial cities along the western Pacific Coast. Tacoma has first secured these trade relations. Such relations are very tenacious.

SEATTLE.
A Town of Marvelous Growth.
"The Queen City of Puget Sound."

Seattle is the county-seat of King County, and is known far and near as the "Queen City of Puget Sound."

It has a present population of 20,000 against 3,500 in 1880. The city contains three national and several private banks; four daily and several weekly journals; one mortgage, loan and trust company; twenty churches; five public school buildings, two of which cost $30,000 and $42,000 each; a territorial university; two private colleges and a girl's academy; besides numerous private schools, three hospitals and an orphan's home. The wholesale and retail stores are too many to enumerate, some of the former doing a business annually of $500,000 to $1,000,000 each. The city is admirably supplied with pure water, both by numerous private companies on a small scale and by the mammoth works of the Spring Hill Water Company, located at Lake Washington. This company has completed a great reservoir on Central Hill, 315 feet above tide level. Connected with it in the city are over fifty-five hydrants, from which five extinguishing streams are thrown far above the highest buildings in the business part of the city. This city has a splendid system of gasworks, also two electric light companies. Both arc and incandescent lights illuminate our streets. Two lines of street cars are in operation and steadily extending outward, and several other lines are projected. It contains more than forty benevolent societies and fraternal lodges; also four well-drilled and equipped militia companies. During the past few years there have been added to its municipal improvements twenty-five miles of graded streets and sixty miles of sidewalk. Some of the recent steps in the progress of Seattle as a metropolis are here given: On October 1, 1887, the free postal delivery system went into effect in the city. A few weeks later Seattle was made the terminus and centre of distribution for all the mails for the entire Puget Sound country; in consequence it has become the central headquarters and home port for destination and departure of the steamboat system of the Sound. Within its maritime jurisdiction are now plying more than eighty steamers. On December 1, 1887, the United States District Land Office was removed to Seattle, making this city the principal seat of the public land business in Western Washington. Arrangements are perfected for two new lines of cable road for street cars to run from the bay back to Lake Washington, and they will soon be laid, namely, as soon as necessary improvements, now in progress, on Madison, Columbia, Mill and Jackson Streets, are completed.

The city of Seattle contains ten saw mills, whose plants cost $4,000,000, which employ over seven hundred men; and also has tributary to it, within a radius of thirty five miles, the mammoth lumbering establishments of Port Blakely, Port Madison, Port Discovery, Port Gamble, Port Ludlow, Utsalady and Seabeck, said to be the largest saw mills in the world, some of them having a capacity of 350,000 feet per diem, and employing scores of sea going ships. There are three or four brick yards and tile factories, four breweries, numerous bakeries, candy factories, a cracker factory, several sash, door and blind factories, shingle factory, soap works, furniture factory, soda works, bottling establishments, carpet weavers, match factory, harness and saddlery, blank books and bindery, book printing, several boiler works, foundries, iron and brass works, etc.; numerous boot and shoe shops and tailoring establishments, factories of shirts and underwear, cigars, millinery goods, chair stock, barrels, plaster decorations, etc.; four marble and stone cutting works, patent medicines, dressmakers, hair work, carriage makers, wagon shops, fish packers, coffee and spice mill, cabinetmakers, boat builders; and numerous dentists, jewelers, watchmakers, florists, nurserymen, fancy poultry breeders and stockmen, furriers, gun and locksmiths. hatters, meat packers, photographers, picture framers and painters, metallic roof works, scroll saw works, shipyards, tin shops, taxidermists, chemists, undertakers, etc.

The export trade of Seattle and Puget Sound is very large and is rapidly increasing. An idea of it can be formed from a single fact. During the fiscal year ending June 30, 1887, the United States Custom House at Port Townsend noted the departure from the Sound of 641 cargoes of coal and lumber, besides several of wheat, which, at $10,000 each (a low estimate), would be worth $6,500,000. A large bulk of this export wealth went to foreign ports all over the world, to be paid for in coin. As Seattle is the chief metropolis of the entire Puget Sound region, it is not far out of the way to credit the most of this business as her commerce, since it is largely contributory to her growth. During the fiscal year ending June 30, 1885, the number of vessels entered in the Puget Sound district was 1,065, with a tonnage of 478,000, and the clearances were 1,065, with a tonnage of 452,234. Of the entrances, 271 cargoes, 151,301 tons were in cargo, and 794 cargoes, 326,839 tons were in ballast. Of the departures, those proportions were just reversed, showing the balance of trade. The total value of her foreign and coastwise exports for 1885 was $7,000,000. Besides the ordinary shipments of coal, lumber, hops, oats, wheat, potatoes, furs, lime, canned and barreled salmon, the daily routine export trade to the neighboring British ports of Victoria and British Columbia forms an enormous item.

Advantages of Seattle. The special advantages of Seattle are too numerous to mention in full. A few may be specified, as: First—A splendid harbor, scarcely equaled in the world for the varied purposes and convenience of commerce. Second—Its central position relative to the commerce of the world, as the great seaport on the Pacific Ocean of North America, and directly facing the teeming population of Asia and the great and rich islands of the South Seas. It is already the chief port of supply for the growing trade of Alaska— a great region, more extensive than the thirteen original States of the Union, with an ocean coast line of thousands of miles, that is beginning now to loom up as a great coming source of supply of the precious metals, as well as of furs, fish, whale oil, yellow cedar and ice. Third—It has an excellent and most

productive soil for fruits, flowers, and garden produce, of such a nature as not to be very dusty in summer nor muddy in winter. Fourth—Its exceptional healthfulness. The death rate in Seattle is only 7 in 1000, per annum, which is less than one-third that of the northern cities of the Union. Fifth—Its mild, even and delicious climate, free from all dangers from the clouds above, from vapors or miasma around, or the fires beneath. Sixth—Its surroundings on all sides, except the magnificent harbor front, by grand lakes and deep, navigable rivers, which have caused it to be officially designated as the location of a great naval station and construction yard. Seventh—The one-third mile canal now completed between

Lakes Union and Washington, in the suburbs of the city, furnishes a great water-power of incalculable value for manufacturing and motive power.

Seattle has two lines of local railroad completed and in operation, the Columbia & Puget Sound, with two branches, one twenty miles long, running to Newcastle, the other forty miles long, running to the Black Diamond and Franklin collieries; and the Puget Sound Shore Line, extending through a link of the Northern Pacific Railroad, and placing the city in connection with the Northern Pacific, the Oregon Railway & Navigation Company, the Union Pacific, the Oregon & California, the Southern Pacific, and the general railroad system of the United States.

The Canadian Pacific has been pushed through the Canadian Dominion by British capital, to a Pacific terminus, something over a hundred miles north of Seattle, and the Seattle & West Coast Railroad, which furnishes the connecting link, and makes Seattle the American terminus of this great system, is now under contract for the entire distance, and is being rapidly pushed to completion this year. This line, as regards the carrying trade, is as much an element in the transportation problem of the Northwest as any of the American roads. The Seattle, Bellingham Bay & British Columbia Railroad Company, a local company of Seattle capitalists, has obtained from Congress a charter for a through line, and are actively pushing the preliminary work for a second line to connect Seattle by rail, direct with the Canadian system, at the international boundary line on the 49th parallel.

Beauty of the City. The city presents a beautiful and striking appearance from whatever side it is approached. It rises from the water front to the crest of a hill in a gradual slope. The site is most beautiful. The city extends about four miles along the water front. The whole water front is lined with mills, manufacturing establishments of various kinds, commission and storage, and warehouses.

Steamers are constantly arriving and departing; regular lines run to Tacoma and Olympia, to Port Townsend and Victoria, to Whatcom and other points on Bellingham Bay, and to the Skagit River; there are regular steamers to Alaska, San Francisco, San Diego, and other points in California. Ships from China, Japan, Australia, crowd its docks. In addition to the great and varied industries on the water front, there are business blocks, higher up, that would do credit to any Eastern city. The residence portion of Seattle is unsurpassed for beauty. There are hundreds of homes costing from $3,000 to $50,000, surrounded by charming grounds, and so located and constructed as to command magnificent views of the Sound, the Olympic and Cascade Ranges of mountains, always covered with snow, and the mighty peaks of Mounts Rainier and Baker. To the north of the city and close up to it lies the beautiful Lake Union, a body of fresh water covering a section or two of land, and of immense depth. The heights about this lake are being covered with pleasant homes, and in the near future it will be a most delightful resort. To the east of the city, four miles from the bay, but now hardly a mile from the city limits, lies Lake Washington, twenty-five miles in length by from two to four in width. It is clear, fresh, sparkling water, so deep that it can not or has not yet been sounded. The lake is hemmed in by hills covered with giant forest trees. The water supply of Seattle is drawn from this lake. It is connected with Lake Union by a small stream, which is being enlarged into a ship canal, so that within a year or two the largest steamers and ships will go directly from the salt water of the Sound into the clear, fresh water of Lake Washington. It will make one of the finest ship-building points and dry-dock stations in the world, and will certainly be utilized for such purposes, either by the National Government or private enterprise. There is certainly not within the National domain such an eligible location for a great navy yard. Special attention is being paid to the establishment of manufacturing industries in Seattle, and almost every week some new enterprise is materialized. Henry Villard, in his visit to the city in 1878, designated it "The Queen City." Situated as it is, in the heart of Western Washington, with railways running out in many directions, with a harbor equal to any in the world, the city well deserves the title. The city is the nucleus of

MT. RAINIER, W. T.

territorial commerce; all the prosperity of the country is reflected in the general progress of the city. The history of the city is the history of the whole Northwest. It is the supply depot and shipping port for a quarter of a million people; it is the wholesale and retail market for a vast territory. Its commerce within the last two years has assumed enormous proportions. It is the coal and lumber shipping depot for the whole Pacific Coast. It is the heart of navigation of Puget Sound. Nearly two hundred steamers radiate from the wharves to different local points. (Population, 20,000. Distance from San Francisco, 940 miles.)

COMPLETE INDEX TO STATIONS ON DENVER & RIO GRANDE RAILROAD AND DENVER & RIO GRANDE WESTERN RAILWAY.

STATIONS	State or Territory	Population	Elevation	Dist. from Denver	STATIONS	State or Territory	Population	Elevation	Dist. from Denver
Acheron	Utah		4638	463	Crested Butte	Colo	1200	8878	312
Acequia	Colo		5530	17	Crevasse	Colo		4526	446
Adobe	Colo			148	Crookton	Colo		8168	267
Alamosa	Colo	1200	7546	250	Crystal Creek	Colo		6831	329
Alcalde	N. M.		5709	359	Crystal Lake	Colo		9332	269
Allenton	Colo		7144	320	Cuchara Junc	Colo	25	5942	169
Almont	Colo		8042	301	Cumbres	Colo		10115	329
Alta	Utah			742	Curecanti	Colo		7075	322
Amargo	N. M.	125	7009	365	Dallas	Colo		6926	376
Americus	Colo		8140	246	Davenport	Colo		8179	241
American Fork	Utah	1800	4567	702	Deer Run	Colo			404
Anth. C'l Mine	Colo		8947	322	Delta	Colo	400	4980	374
Antonito	Colo	250	7888	279	Del Norte	Colo	1200	7880	281
Apache	Colo		5946	164	Denver	Colo	125000	5195	
Apishapa	Colo		6158	189	Desert Switch	Utah		4504	558
Arioles	Colo	25	6013	402	Dillon	Colo	200	8861	313
Aspen	Colo	6500	7868	408	Dotsero	Colo		6154	349
Azotea	N. M.		7728	358	Doyle	Colo		8062	271
Baldy	Colo		7614	238	Douglas	Colo		6323	35
Barnes	Colo		6232	195	Dominges	Colo		4801	392
Barnum	Colo		7793	331	Draper	Utah		4394	776
Barranca	N. M.	25	6949	344	Dulce	N. M.		6779	372
Battle Creek	Utah		4497	698	Duncan	Colo		4880	382
Beaver	Colo		4999	143	Dundee	Colo		4712	117
Belleview	Colo			220	Durango	Colo	3500	6520	450
Bessemer	Colo		4774	121	Eagle Junc	Colo		9762	275
Big Horn	N. M.		9022	298	Eagle Park	Colo		9227	294
Bingham Junc	Utah	500	4366	723	Eagle River	Colo		6598	335
Bingham	Utah	900	4875	740	Echo	Colo		6085	181
Bird's Eye	Colo		10183	232	Edgerton	Colo		6412	55
Blackburn	Colo		7379	183	Elfer	Colo		9858	275
Blanca	Colo		9064	207	Eldredge	Colo		6541	365
Boaz	Colo			186	Elko	Colo			264
Bocea	Colo		6709	444	Elk Park	Colo		8683	489
Bonita	Colo			272	El Moro	Colo	200	5579	206
Borst	Colo		6811	58	Embudo	N. M.	25	5821	351
Bridge 3	Colo		5048	146	Emma	Colo		6610	368
Bridgeport	Colo		4755	399	Engleville	Colo		6493	213
Brown's Canon	Colo		7322	224	Escalante	Colo		4845	386
Buena Vista	Colo	1800	7970	242	Espanola	N. M.	100	5590	370
Burnham	Colo		5241	2	Excelsior	Colo		4928	457
Buxton	Colo		8794	254	Fairy Glen	Colo			179
Buttes	Colo		5636	94	Fairview	Colo			348
Cactus	Colo		4880	112	Farnham	Utah		1534	509
Caliente	N. M.		6324	335	Farmington	Utah		4336	750
Calumet	Colo		9027	232	Florence	Colo	1000	5199	152
Carbon	Colo		8424	448	Ft. Worth Junc	Colo		4805	122
Carbondale	Colo	500	6181	378	Fort Crawford	Colo		6182	361
Carlile	Colo		4950	140	Florida	Colo		6717	436
Canon City	Colo	2500	5343	161	Fountain	Colo	200	5568	68
Carracas	Colo		6173	394	Francklyn	Utah		4291	728
Cascade	Colo		7785	477	Fremont Pass	Colo		11328	290
Castle Gate	Utah		6257	622	Frisco	Colo		9086	310
Castle Rock	Colo	300	6219	33	Fruitvale	Colo	25	4523	436
Cattle Creek	Colo		6037	374	Garland	Colo	100	7936	226
Cebolla	Colo		7354	319	Garfield	Colo	100	9510	235
Cedar Creek	Colo		6755	343	Glaciers	Colo			312
Cerro Summit	Colo		7968	336	Glenwood Spgs	Colo	3000	5768	367
Chama	N. M.	250	7865	343	Germania	Utah		4296	723
Chamita	N. M.		5641	365	Glade	Colo		6518	38
Chester	Colo		9412	250	Goodnight	Colo		4728	124
Chicosa Junc	Colo		6124	199	Gorge	Colo			165
Chipeta	Colo			369	Govetown	Colo		7639	188
Cisco	Utah		4447	490	Graneros	Colo		5804	146
Cimarron	Colo	150	6906	331	Grand Junc	Colo	1500	4594	425
Clear Creek	Utah	50	6228	650	Granite	Colo	100	8945	259
Cleora	Colo		7014	214	Grassy Trail	Utah		4874	581
Coal Mine	Utah			653	Gray's	Colo		9673	236
Coal Creek	Colo	1500	5360	155	Greenborn	Colo		5102	134
Colorado City	Colo	1800	6110	78	Greenland	Colo	25	6921	47
Colorado Spr's	Colo	10000	5982	75	Green River	Utah		4069	544
Colorow	Colo		5352	364	Gulf Junction	Colo		4681	119
Conchita Junc	Colo		6395	181	Gunnison	Colo	2500	7683	290
Coxo	Colo		9753	331	Gypsum	Colo		6325	342
Cotopaxi	Colo	50	6385	193	Hale	Utah			646
Cottonwood	Utah		4602	479	Halfway	Colo			232
Crane Park	Colo		10119	281	Hayes	Colo			247
Crescent	Utah		4896	521	Hayden	Colo		9158	265
Cresco	N. M.		9193	334	Hecla Junc	Colo		7371	226

INDEX TO STATIONS—Continued.

STATIONS.	State or Territory	Population	Elevation	Dist. from Denver	STATIONS	State or Territory	Population	Elevation	Dist. from Denver
Henry	Colo			255	Pike View	Colo		6188	71
Hermosa	Colo		6645	461	Pine Creek	Colo			254
Hilden	Colo		10277	245	Pinon	Colo		5038	106
Home Ranch	Colo		6559	456	Placer	Colo	75	8410	212
Hooper	Utah		4391	754	Plateau	Colo			29
Hotchkiss	Colo		6409	365	Pleasant Val Jc	Utah	200	7177	636
Hot Springs	Colo..r.		9024	255	Pocono	Colo		10316	239
Howard's	Colo		6714	205	Pole Canon	Utah		4890	674
Huerfano	Colo		5677	157	Poncha Junc	Colo	200	7480	221
Husted	Colo		8596	62	Poncha Pass	Colo		9059	231
Ignatio	Colo		6437	424	Price	Utah	100	5547	611
Jack's Cabin	Colo		8309	306	Provo	Utah	5000	4517	689
Jordon Nar'ws	Utah			724	Pueblo	Colo	25000	4667	120
Juanita	Colo		6341	385	Quarry	Colo		6228	79
Kahnah	Colo		4683	409	Red Cliff	Colo	1000	8671	299
Kaysville	Utah		4263	754	Red Narrows	Utah		5543	663
Keeldar	Colo		9970	279	Reno	Colo		5236	157
Keene	Colo		9301	233	Ridgway	Colo			296
Kelker	Colo			80	Riverside	Colo	140	6372	249
Kezar	Colo		7434	302	Roan	Colo		4542	433
Kokomo	Colo	50	10631	296	Robinson	Colo	50	10871	294
Kyune	Utah		7052	632	Rockdale	Colo			190
La Boca	Colo		6177	416	Rockwood	Colo		7367	468
Lake City	Colo	1500	8604	352	Roswell	Colo			73
Lake Hughes	Colo		7470	385	Roulldeau	Colo			379
La Jara	Colo	350	7609	265	Round Hill	Colo		8687	235
Lake Shore	Utah			748	Rouse	Colo		6486	183
Larimer	Colo			150	Rouse Junction	Colo		6144	173
Larkspur	Colo	25	6669	49	Rock Creek	Colo		8304	303
Layton	Utah			787	Russell	Colo			325
Lava	N. M		6466	289	Sagers	Utah			302
La Veta	Colo	300	7025	191	Salida	Colo	3000	7049	317
Leadville	Colo	20000	10200	277	Salt Creek	Colo		5469	140
Lehigh	Utah	2000	4544	705	Salt Lake City	Utah	25000	4228	735
Lehigh Junc	Colo		5694	21	San Carlos	Colo		4900	128
Leon	Colo			386	Sandy	Utah	500		725
Little Grand	Utah		4604	527	S'nta Cl'raMine	Colo		6447	187
Littleton	Colo	300	5372	10	Sapinero	Colo	48	7255	316
Lobato	N. M		8303	339	Sargent	Colo	200	8477	259
Los Pinos	Colo		9637	321	Schofield	Utah	500		652
Lower Crossing	Utah	25	4630	570	Sedalia	Colo	100	5835	25
Malta	Colo	50	9580	272	Servietta	N. M		7727	323
Manitou	Colo	1300	6318	80	Shale	Colo		4608	452
Marshall Pass	Colo		10856	242	Sherwood	Colo		6901	328
Marsh	Colo		6347	171	Shawano	Colo			246
Maysville	Colo	100	8320	228	Sheridan Park	Colo		5316	9
Meadows	Colo		4812	130	Shirley	Colo		8869	230
Mear's Junc	Colo	25	8431	227	Siding No. 1	Colo			363
Menoken	Colo			359	Shoshone	Colo		6119	357
Mesa	Colo			124	Silla	Colo		6672	430
Minturn	Colo		7823	308	Silver Lake	Colo		6395	181
Midway	Colo		7852	239	Silverton	Colo	2500	9224	495
Military Junc	Colo		5329	8	Soda Springs	Colo		6850	176
Military Post	Colo		5393	10	Soldier Sum'it	Utah		7465	642
Mill Fork	Utah		5806	658	Solitude	Utah		4283	507
Mitchell's	Colo		9922	287	South Fork	Colo		8188	297
Monarch	Colo	500	10028	237	Spanish Fork	Utah	2500	4721	679
Monero	N. M	100	7264	362	Sphinx	Utah			551
Montrose	Colo	1500	5311	353	Spikebuck	Colo			178
Monte Vista	Colo	1000	7665	267	Spring Creek	Utah			616
Monument	Colo	200	6974	56	Springville	Utah	2500	4565	683
Mounds	Colo			284	State Line	Colo		4753	461
Mule Shoe	Colo		8754	202	Stewart Junc	Colo		8006	317
Nathrop	Colo	50	7695	234	Sublette	N. M		9276	305
Navajo	N. M		6588	376	Sunnyside	Utah		5270	591
Needleton	Colo		8141	481	Swallows	Colo		4878	135
New Castle	Colo	300	5560	379	Swissvale	Colo			209
No Agua	N. M		8205	306	Tennessee Pass	Colo	25	10433	283
Oak Creek	Colo		5352	156	Texas Creek	Colo	52	6217	186
Ogden	Utah	10000	4286	771	Thistle	Utah	100	5043	679
Ojo	Colo		6519	199	Thompson	Utah		5145	515
Otto	Colo		8212	228	Tioga	Colo		6203	160
Osler	Colo		9637	317	Tollgate	Colo			166
Ouray	Colo	2500	7721	389	Toltec	N. M		9465	309
Ouray Junc	Colo		5830	354	Toluca	Colo			21
Overland Park	Colo		5275	6	Tres Piedras	N. M	200	8088	313
Palmer Lake	Colo	150	7237	52	TrimbleSpr'gs	Colo		6650	459
Palmilla	N. M		8258	290	Trinchers	Colo		8104	220
Parkdale	Colo	30	5737	171	Trinidad	Colo	6000	5994	210
Parlin's	Colo	100	7952	278	Tuna	Colo			174
Parma	Colo		7616	261	Twin Lakes	Colo		9027	261
Petersburg	Colo	50	5322	8	Unaweep	Colo		4636	416
Picton	Colo		6265	180	Vallejo	Colo		6222	409
Piedmont	Colo		7108	382	Valle	Colo		6534	199

INDEX TO STATIONS—Continued.

STATIONS.	State or Territory	Population	Elevation	Dist. from Denver	STATIONS.	State or Territory	Population	Elevation	Dist. from Denver
Vegas	Colo.			127	Westwater	Utah			473
Veta Pass	Colo.		9393	206	Wheeler	Colo.		9781	302
Villa Grove	Colo.	200	7971	247	Whitehouse	Utah		4486	499
Volcano	N. M.		8487	297	White Water	Colo.		4645	413
W'g'n Wh'lGap	Colo.	25	8449	311	Widefield	Colo.		5720	84
Wahatoya	Colo.		8504	183	Wigwam	Colo.		5231	99
Walsen's	Colo.	1000	6189	176	Willow Creek	N. M.		7742	348
Wasatch	Utah	25		734	Wood's Cross	Utah	100	4255	740
West Cliff	Colo.	800	7864	191	Woody Creek	Colo.		7270	399
West Denver	Colo.		5201	1					

MOUNTAIN PEAKS AND PASSES OF COLORADO.

With their elevations above sea level.

	Feet.		Feet.		Feet.
Blanca	14,464	Holy Cross	14,176	Pigeon	13,928
Harvard	14,383	Baldy	14,176	Blanc	13,905
Massive	14,368	Sneffles	14,158	Frustrum	13,888
Gray's	14,341	Pikes	14,147	Pyramid	13,885
Rosalie	14,340	Castle	14,106	White Rock	13,847
Torrey	14,336	Yale	14,101	Hague	13,832
Elbert	14,326	San Luis	14,100	R. G. Pyramid	13,773
La Plata	14,302	Red Cloud	14,092	Silver Heels	13,766
Lincoln	14,297	Wetterhorn	14,069	Hunchback	13,755
Buckskin	14,296	Simpson	14,055	Rowter	13,750
Wilson	14,280	Aeolus	14,054	Homestake	13,687
Long's	14,271	Ouray	14,043	Ojo	13,640
Quandary	14,269	Stewart	14,032	Spanish	13,620-12,720
Antero	14,245	Maroon	14,000	Guyot	13,565
James'	14,242	Cameron	14,000	Trinchara	13,546
Shavano	14,238	Handie	13,997	Kendall	13,542
Uncompahgre	14,235	Capitol	13,992	Buffalo	13,541
Crestones	14,233	Horseshoe	13,988	Arapahoe	13,520
Princeton	14,199	Snowmass	13,961	Dunn	13,502
Mount Bross	14,185	Grizzly	13,956	Bellevue	11,000

MOUNTAIN PASSES.

	Feet.		Feet.		Feet.
Alpine Pass	13,550	Berthoud Pass	11,349	Breckenridge Pass	9,496
Argentine Pass	13,100	Marshall Pass	10,852	Cottonwood Pass	13,506
Cochetopa Pass	10,032	Veta Pass	9,392	Fremont Pass	11,540
Hayden Pass	10,780	Poncha Pass	8,945	Mosquito Pass	13,700
Trout Creek Pass	9,346	Tennessee Pass	10,418	Ute Pass	11,200
		Tarryall Pass	12,176		

Seventy-two peaks between 13,500 and 14,300 feet in height are unnamed and not in this list.

ELEVATION OF LAKES.

	Feet.		Feet.		Feet.
Twin Lakes	9,357	Chicago Lakes	11,500	Palmer Lake	7,236
Grand Lake	8,153	Evergreen Lakes	10,500	Cottonwood Lake	7,700
Green Lakes	10,000	Seven Lakes	11,806		

ALTITUDES OF TOWNS AND CITIES.

REVISED SINCE FIRST EDITION FROM ENGINEERS' MEASUREMENTS.

Town	Feet	Town	Feet	Town	Feet
Alamosa	7,546	El Moro	5,879	Ogden, Utah	4,286
Animas City	6,554	Ft. Garland	7,936	Pagosa Springs	7,108
Animas Forks	11,200	Granite	8,945	Pinos, Chama Summit	9,902
Antonito	7,888	Grand Junction	4,583	Poncha Springs	7,460
Aspen	7,775	Gunnison	7,680	Palmer Lake	7,236
Buena Vista	7,970	Glenwood Springs	5,200	Pueblo	4,669
Canon City	5,344	Howardsville	9,700	Red Cliff	8,671
Castle Rock	6,220	Irwin	10,500	Robinson	10,871
Colorado Springs	5,992	Kokomo	10,631	Rosita	8,500
Crested Butte	8,875	Lake City	8,550	Ruby Camp	10,500
Conejos	7,880	La Veta	7,024	Saguache	7,723
Cottonwood Springs	7,950	Leadville	10,200	Salt Lake City	4,228
Cuchara	5,943	Los Pinos	9,637	Silver Cliff	7,816
Cumbres	10,015	Montrose	5,793	Silverton	9,224
Delta	4,963	Malta	9,580	Salida	7,050
Del Norte	7,880	Manitou	6,324	Trimble Springs	6,644
Denver	5,196	Ojo Caliente	7,324	Westcliffe	7,864
Durango	6,520	Ouray	7,640	Wagon Wheel Gap	8,448

DISTANCES FROM DENVER.

Place	Miles	Place	Miles	Place	Miles
Alamosa	250	Estes Park	85	Parrott City	466
Animas City	390	Eureka, Neb.	394	Pueblo	118
Antonito	278	Fort Garland	226	Philadelphia, Pa.	1,888
Albany, N. Y.	1,920	Grand Junction	425	Pittsburgh, Pa.	1,586
Breckenridge	198	Granite	259	Peoria, Ill.	979
Buena Vista	135	Gunnison	290	Quincy, Ill.	845
Boston, Mass.	2,121	Huerfano	157	Rosita	190
Baltimore, Md.	1,657	Irwin	230	Red Cliff	245
Buffalo, N. Y.	1,652	Indianapolis, Ind.	1,166	Rock Island, Ill.	1,024
Canon City	160	Kokomo	297	Saguache	260
Colorado Springs	75	Kansas City, Mo.	639	Silverton	495
Conejos	270	Lake City	364	Salida	245
Cucharas	170	LaVeta	191	Silver Cliff	194
Castle Rock	33	Leadville	171	South Pueblo	120
Cincinnati, O.	1,255	Littleton	11	St. Louis, Mo	913
Chicago, Ill	1,059	Louisville, Ky	1,233	San Francisco, Cal.	1,445
Cleveland, O.	1,469	Las Vegas	341	Salt Lake City, Utah	660
Columbus, O.	1,341	Manitou	80	Santa Fe, N. M.	800
Columbus, Ky.	1,108	Monument	56	Trinidad	211
Del Norte	285	Montrose	353	Trimble Springs	459
Dillon	313	Milwaukee, Wis	1,197	Twin Lakes	261
Deadwood, M. T.	387	New York, N. Y.	1,910	Toledo, O.	1,303
Durango	450	Nashville, Tenn	1,418	Veta Pass	203
Detroit, Mich.	1,343	Ouray	425	Walsen's	176
El Moro	206	Omaha, Neb.	569	Wagon Wheel Gap	311
Espanola	370	Ogden, Utah	622	Washington, D. C.	1,809
		Palmer Lake	52		

PRONUNCIATION OF PROPER NAMES.

Acequia	A-sa kia	Ojo Caliente	O-ho Cal-i-en-te
Crested Butte	Crested Bute	Ojo	O-ho
Costilla	Costea	Pueblo de Taos	Pueblo-de-Tows
Canon	Can-yon	Pinon	Pin-yon
Cumbres	Cum-breez	Saguache	Se-watch
Cuchara	Cu-cha-ra	Sierra Mojeda	Sierra Mo-ya-da
Conejos	Co-na-jos	Santa Fe	San-ta Fay
Chihuahua	Che-wa-wa	San Juan	San Wan
Huerfano	Wa-far-no	San Miguel	San-me-gil
La Junta	La Hun-ta	Sapinero	Sapi-na-ro
La Jara	La Hara	Tierra Amarilla	Tier Ama-rea
La Veta	La Va-ta	Trinchera	Trin-chara
Monero	Mo-na-ro	Vallejo	Vall-a-ho
Manitou	Man-i-too	Wahatoya	Wa-ha-toy-ya
Navajo	Na-va-ho		

COMPLETE INDEX TO STATIONS ON THE SOUTHERN PACIFIC RAILROAD.

STATIONS	State or Terr'ty	Population	Elevation	Miles from S.Francisco	STATIONS	State or Terr'ty	Population	Elevation	Miles from S.Francisco
Acampo	Cal	120	59	107.21	Blue Canon	Cal	110	4695	168
Acton	Cal	150	2670	427.1	Blue Creek	Utah		4272	792
Aden	N.M.		4391	1237.9	Boca	Cal	400	5531	218
Adonde	Ariz.		212	760.9	Bolsa	Cal		177	89
Afton	N.M.		4207	1250	Bonneville	Utah		4260	823
Ager	Cal	100	2346	387	Boyce's	Cal			108.29
Agnews	Cal	60	25	41	Borden	Cal	386	273	187.81
Alameda	Cal	7750	20	11.13	Bosque	Ariz		1080	859.5
Alamitos	Cal		191	58	Boulder Creek	Cal	300	470	81
Albany	Or	3000	210	692	Bovine	Utah		4277	699
Alcalde	Cal		850	301.34	Howie	Ariz	100	3759	1089
Alder Creek	Cal	50		109	Bracks	Cal			117.29
Alexis	Cal			558.2	Bradley	Cal		539	196
Alhambra	Cal	700		490	Brandon	Cal			130
Alila	Cal	35	280	273.51	Brentwood	Cal	200	80	62.70
Allendale	Cal			71.22	Brighton	Cal	500	54	134.03
Alma	Cal	170	560	58	Brigham	Utah	1800	4239	816
Almaden	Cal	1500	348	53	Brooks	Or	35		728
Almond	Cal	60	92	515.3	Brookside	Cal			546.6
Alpine	Cal		2932	417.3	Browns	Cal		3929	325
Alta	Cal	50	3607	159	Bryant	Cal			129
Altamont	Cal	40	740	55.97	Buckeye	Cal			301
Alvarado	Cal	600	15	24	Buckhorn	Cal		593	468.5
Alviso	Cal	110	8	38	Burbank	Cal	250	558	471.4
Amaranth	Cal			105.17	Burnetts	Cal		189	123.70
Am. Rvr Bridg.	Cal		49	93	Burson	Cal	30		126.36
Amity	Or	225		890	Butler	Cal			212.95
Anaheim	Cal	2000	133	508.9	Byron	Cal	100	34	67.83
Anderson	Cal	750	432	249	Cabazon	Cal	100	1779	574.9
Anita	Cal		160	195	Cachise	Ariz		4222	1054.5
Annadel	Cal			67.91	Cactus	Cal		395	712.8
Antelope	Cal	100	162	104	Cadanassa	Cal			97.47
Antioch	Cal	700	46	54.54	Calicate	Cal	50	1290	336.3
Applegate	Cal		2014	136	Calistoga	Cal	850	363	72.79
Aptos	Cal	100	102	112	Campbell	Cal		65	51
Araby	Ariz		144	736.6	Cambray	N.M.		4324	1224.5
Arbuckle	Cal	300	139	114.47	Cameron	Cal		3787	30.9
Arcaie	Cal		55	98	Camilos	Cal	150	733	463.1
Arena	Cal		141	140.09	Cana	Cal	200	172	196
Argenta	Nev		4547	488	Canby	Or	100		749
Arimona	Cal		238	257.43	Canemat	Or			756
Army Point	Cal		11	31.86	Cannon	Cal		92	56.35
Ashland	Or	2000	1898	431	Cantara	Cal			330
Athena	Cal			161.57	Capay	Cal	200		98 37
Athlone	Cal	50	210	161.64	C. H. Mills	Cal		2676	149
Atwater	Cal	20	153	141.28	Carbondale	Cal	75	221	133.22
Auburn	Cal	1700	1360	126	Carlin	Nev	394	4897	635
Aurora	Or	200		745	Carlton	Or	125		675
Aurant	Cal			485	Carmenita	Cal		74	501.1
Avon	Cal		12	39.15	Carnadero	Cal		168	82
Aztec	Ariz		495	805 3	Carpinteria	Cal	300	8	517.3
Baden	Cal		39	12	Casa Grande	Ariz	400	1396	913.7
Bakersheld	Cal	2000	415	314.04	Cascade	Cal	28	6538	190
Bale	Cal		287	68.53	Cashmere	Cal			108.77
Balfour	Utah		4239	801	Castale	Cal		1004	454 3
Banning	Cal	200	2317	569	Castle	Cal		37	97.59
Santa	Cal	150	30	74.82	Castle Crag	Cal		1943	317
Bardins	Cal		48	115	Castroville	Cal	600	17	110
Barlaws	Or			747	Cemetery	Cal		92	1
Barrett	Cal		39	16	Cedar	Nev		5974	620
Barro	Cal		245	66.35	Centerville	Cal	300		32
Batavia	Cal	30	27	65.15	Central Point	Or	50		450
Battle Mount'n	Nev	522	4511	474	Ceres	Cal	200	93	118.65
Bay Point	Cal		10	42.24	Cerritos	Cal		36	498.5
Bealville	Cal		1793	341.6	Charleston	Cal		40	98.20
Beaumont	Cal	300	2560	562.8	Chemawa	Or			724
Beaverton	Or	250			Chestnut	Cal			322
Bello	Cal		204	62.32	Chittendens	Cal			91
Belmont	Cal	202	31	25	Chico	Cal	6005	193	186
Benicia	Cal	3200	10	33.25	Choione	Cal		232	152
Ben Lomond	Cal			77	Chromite	Cal		1541	309
Bennett	Cal			134	Chuslar	Cal	75	103	129
Benson	Ariz	2000	3578	1035.4	Cicero	Cal		90	121.45
Beowawe	Nev	82	4895	507	Cienega	Cal	100	110	490
Berenda	Cal	85	258	177.59	Cisco	Cal	25	5931	182
Bernal	Cal		186	4	Citronia	Cal			64.34
Berryman	Cal	5000	10	11.86	Clackamas	Or	150		762
Bethany	Cal	103	40	76.71	Clarks	Nev		5263	264
Biggs	Cal	1000	98	183	Clawson	Or		2250	425
Big Trees	Cal	250	270	74	Clements	Cal			115.30
Bishops	Nev		5423	602	Cluro	Nev		4785	516
Bituma	Cal		12	502.7	Clyde	Cal		153	119.20
Blacks	Cal	100	52	96.74	Coalinga	Cal		665	298.34
Bl'k Butte Smt.	Cal			345	Copper Gap	Cal	50	1759	133

INDEX TO STATIONS—Continued.

STATIONS.	State or Terr'ty	Population	Elevation	Miles from S. Francisco	STATIONS.	State or Terr'ty	Population	Elevation	Miles from S. Francisco
Coburns	Cal		259	158	Emigrant Gap	Cal	20	5221	174
Coin	Nev		4596	462	Esperanza	Cal			90.77
Coles	Cal		2905	404	Essex	Nev		4936	232
Colestin	Or		3730	411	Estrella	Ariz		1521	869
Colfax	Cal	400	2422	144	Eugene	Or	2200	455	649
Colfred	Ariz			778.2	Ewings	Cal	24	120	123
Colma	Cal			9	Exeter	Cal			258.55
Colorado	Cal			171	Fair Grounds	Or			722
Colusa Junct	Cal	5	84	730.5	Fair Oaks	Cal			31
Colton	Cal	2500	965	130.13	Farmington	Cal	350	115	11.2
Cometa	Cal		153	539.7	Felton	Cal	200	275	74
Compton	Cal	800	76	117.20	Felton (old)	Cal		275	76
Comstocks	Or	25		493.7	Fenelon	Nev	100	6154	635
Coopers	Cal		23	618	Fernando	Cal	200	1066	461
Copley	Cal	25	600	113	Filbert	Cal	45		80
Cordelia	Cal	100	15	270	Fillmore	Cal	75	475	473.7
Corinne	Utah	500	4231	45.89	Finnell	Cal		258	184.68
Cornelius	Or	200		809	Florence	Cal	200	151	486.2
Corning	Cal	350	277	857	Florin	Cal	100	42	129.98
Cornwall	Cal	75	30	79.53	Florson	Cal		5353	223
Corvallis	Or	1800		49.89	Flowing Well	Cal		5	677
Cosgrave	Nev		4237	929	Folsom	Cal	1000		112
Cothrin	Cal			393	Fortuna	Cal			226.45
Cottage Grove	Or	300		124	Forest Grove	Or	800		859
Cottonwood	Cal	450	421	626	Forest Lake	Cal		52	110.50
Coyote	Cal	200	251	242	Fowler	Cal	150	308	216.23
Cranor	Or			63	Flosden	Cal			35.39
Creston	Cal		315	698	Fresno	Cal	8000	293	206.65
Creswell	Or	100		41.97	Friuks	Cal		260	653.3
Crocketts	Cal	100	12	635	Froman	Or			693
Cross Creek	Cal		278	30	Fruitvale	Cal	250	33	10.72
Crowleys	Or			234.79	Fruto	Cal			167.97
Cucamonga	Cal	200	952	899	Fry	Or			695
Cummings	Cal			521.5	Fulton Wells	Cal	100	124	499.3
Curtis	Cal		42	141	Gage	N. M.	25	4488	1179
Daulton	Cal		408	91.03	Galt	Cal	700	49	112.6
Dathol	Cal		495	189.17	Gartney	Utah			681
Davis	Cal	500	53	287.94	Gaston	Or	75		865
Decoto	Cal	150	68	76.56	Gazelle	Cal		2760	363
Deeth	Nev	100	5340	27.42	Geneora	Cal	100	99	118.53
Delano	Cal	500	313	594	Germantown	Cal	200	170	157.81
Delavan	Cal		95	281.84	Gervais	Or	300		344
Delhi	Cal		121	139.28	Gibson	Cal		1387	306
Delta	Cal	20	1138	133.19	Gill Bend	Ariz	50	737	850.1
Del Monte	Cal		8	298	Gill City	Ariz		171	744.7
Deming	N. M.	2000	4334	124	Gillespie	Cal			107.3
Derry	Or			1198.5	Gilroy	Cal	2000	193	80
Desert	Nev		4018	905	Girvan	Cal	200		255
Dillard	Or	150		387	Girard	Cal		3301	355.3
Diamond	Cal	40		564	Giants	Cal	30		696.7
Dilleys	Or	150		146	Glenburn	Cal			306.99
Dinuba	Cal			861	Glen Ellen	Cal	100		60.11
Divide	Or			236.25	Gloster	Cal	125		388.3
Dixon	Cal	1350	66	622	Glendale	Or	100	965	510
Domingues	Cal		61	68.50	Glenwood	Cal		890	66
Dos Palmas	Cal		253	496	Golconda	Nev	335	4392	431
Downey	Cal	1000	111	642.4	Gold Hill	Or	65		459
Dragoon Sum'lt	Ariz	4614		494.9	Gold Run	Cal	250	3222	155
Drains	Or	300		1044.5	Goleta	Cal	400	19	537.1
Drummond	Cal			611	Goltra	Or			697
Dryiyn	Cal		396	62.11	Goodyears	Cal		11	39.01
Dugan	Cal			707.9	Goshen	Cal	75	286	241.64
Dunnigan	Cal	109	69	134	Goshen	Or	30		643
Dunsmuir	Cal	350	2285	104.26	Gonzales	Cal	150	137	135
Durham	Cal	200	161	324	Granite Point	Nev		3916	333
Dutch Flat	Cal	500	3593	180	Grants Pass	Or	1000		476
Dry Camp	Cal			157	Greenwood	Cal		232	162.94
East Oakland	Cal	5300	12	602.2	Greens	Or			570
East Portland	Or	8000	53	9.30	Gregory	Or		3462	409
E. San Gabriel	Cal		409	772	Grenada	Cal			371
Eden Vale	Cal		180	492	Greystone	Cal		267	61
Edgewood	Cal	100	2955	57	Gridley	Cal	1000	97	180
El Casco	Cal		1874	355	Guadaloupe	Cal			56
El Dorado	Cal	300		554.2	Guinda	Cal			106.37
Elk Grove	Cal	301	51	143	Hafed	Nev		251	
Elkhorn	Cal			123.84	Haggin	Cal			97
Elko	Nev	752	5065	107	Halconera	Cal		52	159.37
Ellis	Cal		76	558	Halls	Cal	75		25
Elmira	Cal	350	79	69.59	Halleck	Nev	42	5229	581
El Modena	Cal	200	242	60.32	Haisey	Or	350		675
Elmore	Cal	15	805	518.8	Hanford	Cal	800	242	253.84
El Paso	Texas		3713	245	Harrington	Cal		137	109.30
El Verano	Cal	200		1286.7	Harrisburg	Or	500		667
Elwood	Cal		93	54.81	Hartley	Cal		139	69.27
Ely	Cal			541.3	Haywards	Cal	1500	74	21.08
Emerald	Cal			81.19	Heaton	Ariz		1186	687.2

INDEX TO STATIONS—Continued.

STATIONS	State or Terr'ty	Population	Elevation	Miles from S. Francisco	STATIONS	State or Terr'ty	Population	Elevation	Miles from S. Francisco
Henline	Cal		211	283.44	Los Medanos	Cal		38	52
Herbert	Cal		922	196.2	Los Nietos	Cal		156	500.4
Herndon	Cal	100	280	196.8	Lovelocks	Nev	500	3977	341
Highland	Cal		12	12	Lwr Soda Sprgs	Cal	100	2085	320
Hilgirt	Cal			59.21	Lucin	Utah		4496	688
Hillsboro	Or	800			Luzena	Ariz			1060.4
Hillsdale	Cal		147	55	Lyunly	Cal			265.45
Holborn	Nev		5975	629	Lyman	Cal		144	154.24
Holden	Cal		82	102.7	Lynnwood	Cal		89	492.1
Hollister	Cal	2300	284	94	Machine Shops	Or			769
Homestead	Cal			93	Macy	Cal		94	119.46
Honby	Cal		1200	446.3	Madera	Cal	700	278	185.03
Hooker	Cal			235	Madison	Cal	200		87.67
Hopevale	Cal		116	534	Madrone	Cal		342	69
Hornbrook	Cal	100	2154	395	Mall Dock	Cal		11	34.13
Hot Springs	Nev	42	4072	298	Haitland	Cal			488.1
Hubbard	Or	250		740	Malaga	Cal			211.32
Humboldt	Nev	32	4236	374	Malton	Cal		258	170.01
Huron	Cal	150	367	280.74	Mammoth T'nk	Cal		257	683.1
Igerna	Cal		3730	347	Manlove	Cal			98
Independence	Or	800		906	Maple	Cal		319	69.94
Indio	Cal		20	611.7	Marcus	Cal		173	513.8
Ione	Cal	1000	287	139.80	Maricopa	Ariz		1173	592.2
Iron Point	Nev		4375	442	Marion	Or	300		706
Irving	Or	50		655	Martins	Cal		14	113
Irvinville	Or			699	Matlin	Utah		4597	720
Irvington	Cal	500	75	33.75	Martinez	Cal	1500	10	35.64
Ivy	Cal			493.4	Marysville	Cal	6000	66	143
Javnes	Ariz		2241	971.8	Maxwell	Cal	450	94	134.04
Jefferson	Or	250		791	Mayfield	Cal	900	28	35
Junction	Cal	250	163	108	Maybew	Cal			99
Junction City	Or	350		662	McAvoy	Cal		16	45.5
Kaweah	Cal			254.25	McCloud	Cal		3349	335
Keene	Cal		2705	349.9	McConnells	Cal	100	49	120.30
Kelton	Utah	135	4222	743	McCoy's	Or	100		
Kennet	Cal		669	277	McMinnville	Or	1800		
Keyes	Cal		97	121.98	McPherson	Cal	100	262	517.6
Kimberlena	Cal			296.97	Medford	Or	500	1399	446
Kingsburg	Cal	450	300	226.7	Melitta	Cal			70.41
Kings City	Cal	200	332	163	Melrose	Cal	100	18	11.54
Kirkwood	Cal		226	174.35	Menlo Park	Cal	400	64	52
Knights Ldg	Cal	350	45	95.13	Merazo	Cal			45.51
Kolmar	Utah			789	Merced	Cal	3000	171	151.73
Kurand	Cal			156.77	Merlin	Or			445
Lake	Utah		4215	763	Merritts	Cal		56	81.61
Lake View	Cal		6245	198	Mescal	Ariz		4034	1016.8
Lanark	N. M		4165	1259.7	Mesquite	Cal		294	994.1
Lancaster	Cal	75	2350	405.5	Middle Creek	Cal	150	526	283
Lander	Cal		2225	143	Midway	Cal	35	356	53.93
Lang	Cal		1681	439.3	Millers	Or	20		698
La Patera	Cal		4	538.6	MillCity	Nev	100	4226	386
Latham	Or			625	Millbrae	Cal	100	8	17
Lathrop	Cal	600	26	82.82	Milpitas	Cal	500	23	41.72
Latrobe	Cal	150		127	Mills	Cal			102
Laurel	Cal	100	910	64	Millsholm	Cal			160.57
Laurel Creek	Cal		19	24	Milton	Cal	200	381	121.7
Lawrence	Cal	300	64	44	Milwaukee	Or	125		766
Lawton	Nev		4048	259	Minneola	Cal			215.65
Lebanon	Or	500		701	Minturn	Cal		242	168
Le Francs	Cal			56	Miraflores	Cal		138	510.9
Leland	Or	40		496	Mirage	Nev		4247	513
Lemon	Cal			507.4	Modesto	Cal	2500	91	114.34
Lerdo	Cal		414	301.67	Moh'wk Sum'it	Ariz		541	787
Ligurta	Ariz			754.3	Mojave	Cal	150	2751	381.7
Linoore	Cal	400	220	261.84	Moleen	Nev		4981	546
Lillis	Cal		214	267.29	Monson	Cal			240.75
Lincoln	Cal	600	167	119	Montague	Cal		2542	377
Lisbon	N. M		4278	1149.6	Montalvo	Cal		89	495.3
Live Oak	Cal	200	80	153	Monte	Cal	200	286	495.3
Livermore	Cal	250	485	47.88	Montecito	Cal	102	15	525
Livingston	Cal	50	136	137.49	Montello	Nev		4991	687
Luckford	Cal	300		111.31	Monterey	Cal	2300		125
Lodi	Cal	700	55	104.29	Montezuma	Ariz	50	1330	878.1
Logandale	Cal		108	145.30	Monument	Utah		4226	756
Loino	Cal		72	149	Moore	Nev		6167	623
Loomis	Cal	400	400	115	Morley	Cal		722	282
Long Beach	Cal	750	41	508.8	Mornojo	Cal		13	112
Lorenzo	Cal	125	35	18.43	Morrano	Cal		50	99.84
Lordsburg	N. M	300	4245	1138.9	Mott	Cal	200	3156	333
Loray	Nev		5595	657	Mound City	Cal		1055	543.1
Los Angeles	Cal	60000	293	482.2	Mountain View	Cal	400	73	39
Commerial st			278	403.3	Mt. View	Or			
San Pedro st			258	484	Mt. Eden	Cal	350	20	20
Washingtn st			222	485.6	Mowrys	Cal		12	17
Los Gatos	Cal	2000	400	55	Muddy	Ur			670
Los Gulicos	Cal	100		65.61	Mullen	Cal		65	84.12

INDEX TO STATIONS—Continued.

STATIONS	State or Terr'ty	Population	Elevation	Miles from S. Francisco	STATIONS	State or Terr'ty	Population	Elevation	Miles from S. Francisco
Murphys	Cal		95	42	Railroad Pass	Ariz		4394	1073.6
Muscatel	Cal		299	198.69	Rawson	Cal		287	220
Myrtle Creek	Or	50		552	Raymond	Cal	100	938	198.59
Mystic	Cal		5164	227	Red Bluff	Cal	3500	307	225
Nadeau	Cal		150	491	Redding	Cal	2500	551	260
Napa	Cal	6000	20	46.47	Red Rock	Ariz	50	1865	945.6
Napa Junction	Cal	150	79	38.31	Redwood	Cal	1500	9	28
Natchez	Nev		5295	590	Reedley	Cal			231.05
Natoma	Cal			108	Reedville	Or	15		
Nelson	Cal		124	173	Reeds	Cal		72	136
New Almaden (S.P.C. Ry)	Cal	1500		61	Reno	Nev	4302	4497	244
Newark	Cal	600	25	29	Rice Hill	Or	1500	487	574
N. E. Mills	Cal		2280	139	Richfield	Cal		276	182.80
New Era	Or				Riddles	Or	80		546
Newhall	Cal	300		752	Rillito	Ariz		2058	961.3
Newcastle	Cal	125	1265	452.3	Rincon	Cal		300	77
Newman	Cal	350	956	121	Ripon	Cal	200	72	104.56
New Ramona	Cal			96	Rocklin	Cal	800	249	112
Newton	Or			882	Rogers	N. M		3728	1282.2
Nichols	Or			533	Rosamond	Cal		2315	395.5
Niles	Cal	150	88	30.20	Roscoe Spurs	Cal			467
Norman	Cal	80	96	142.94	Roseburg	Or	1500	487	574
Nord	Cal	200	153	193	Rose Creek	Nev		4820	403
North'n Junc	Cal		58	86.62	Routier	Cal			101
North Vallejo			14	32.11	Rowan	Cal			346.4
N. Val'jo Whf		5500	12	31.05	Rozel	Utah		4588	772
North Yamhill	Or	200		872	Ruckles	Or			517
Norwalk	Cal	200	92	499	Rumsey	Cal			111.37
Oakdale	Cal	1000	155	125.70	Russells	Cal			19
Oak Grove	Cal		17	19	Rutherford	Cal	100	183	60.38
Oak Knoll	Cal		114	51.11	Rye Patch	Nev	65	4257	383
Oakland	Cal	55000	12	7.70	Sacramento	Cal	32000	30	89.79
Oakland	Or	400		591	Sacramento	Cal	32000	30	139.67
Oakville	Cal	25	160	58.49	Salem	Or	6000	190	720
Ocean View	Cal	75	293	7	Salida	Cal	50	72	107.58
Ochoa	Ariz	65	4102	1035.1	Salinas	Cal	3000	44	118
Ogden	Utah	15000	4286	833	Salton	Cal		263	636.7
Ogilby	Cal		355	715.2	Salvia	Nev		4177	271
Ombey	Utah		4721	730	San Carlos	Cal		21	26
Olga	Ariz			1097.2	Sanger Junc	Cal			230.85
Ontario	Cal	1800	981	521	Sand Cut	Cal			94
Orange	Cal	679	127	513.8	Sand Cut	Cal			326.2
Oreana	Nev	55	4181	353	San Andros	Cal		153	106
Oregon City	Or	1400	95	757	San Ardo	Cal		452	182
Orland	Cal	500	259	166.4	San Buenaventura	Cal	3000	45	500.2
Ortega	Cal		77	521.3	San Bruno	Cal	50	15	14
Osino	Nev		5134	566	San Francisco	Cal	400000	14	
Paciouma	Cal			463	San Gabriel	Cal	1000	409	491.4
Painted Rock	Ariz		726	834.3	San Joaquin Bridge	Cal		36	90
Pajaro	Cal	50	23	99	San Jose	Cal	25000	80	46
Palisade	Nev	252	4840	525	San Jose	Cal	25000	91	47.74
Pampa	Cal			329	San Jose	Cal	25000	88	50
Pantano	Ariz		3536	1007.1	San Leandro	Cal	1600	48	15.73
Papago	Ariz		3010	993	San Lucas	Cal	200	396	172
Paper Mill	Or			759	Salsbury	Cal			106
Parkers	Or	35		914	San Mateo	Cal	950	22	21
Paso Robles	Cal	600	723	216	San Miguel	Cal	400	616	207
Peko	Nev		5204	578	San Pablo	Cal	250	30	17.59
Penryn	Cal	400	626	118	San Pedro	Cal	500		507.1
Peplin	Utah			734	Sanswain	Cal		1074	531.2
Perkins	Cal	100		97	San Simon	Ariz		3609	1104.7
Perrys	Cal		299	66	Santa Ana	Cal	5000	134	515.8
Peters	Cal	75	100	106.70	Santa Barbara	Cal	7500	3	527.6
Pequap	Nev	40	6181	640	Santa Clara	Cal	3000	75	44
Phoenix	Or	300		439	Santa Clara	Cal	3000	72	47
Picacho	Ariz		1416	931.9	Santa Cruz (S.P.C. Ry.)	Cal	7000	15	80
Pilot Knob	Cal		285	721.5	Santa Cruz Beach	Cal		9	81
Pinole	Cal	250	10	24.02	Santa Cruz	Cal		18	121
Piru	Cal			486.3	Santa Cruz Beach (N.D.)	Cal		9	120
Pixley	Cal	50	262	297.92	Santa Monica	Cal	2000	500	4
Plute	Nev		4507	468	Santa Rosa	Cal	7000	65	75.01
Placerville	Cal	2000		149	Santa Paula	Cal	900	286	483.7
Pleasanton	Cal	600	353	41.80	Sargents	Cal		40	87
Point Isabel	Cal		10	12.77	Saticoy	Cal		95	146	490.8
Pomona	Cal	4200	857	515	Saugus	Cal		1159	449.8
Popes	Cal			109.29	Sauterne	Cal			107.07
Port Costa	Cal	125	12	32.17	Savanna	Cal		150	296	493.2
Porterville	Cal	250		275.95	Seacliff	Cal		11	509.7
Portland	Or	50000		723	Seco	Utah		4223	748
Poso	Cal		417	293.70	Selby	Cal	300	12	28.51
Proberta	Cal			217	Selma	Cal	2200	311	221.76
Promontory	Utah	131	4905	780					
Prosser Creek	Cal		5590	216					
Puente	Cal	500	323	501.5					
Pyramid	N. M		4301	1134.5					

17

INDEX TO STATIONS—Continued.

STATIONS	State or Terr'ty	Population	Elevation	Miles from S. Francisco	STATIONS	State or Terr'ty	Population	Elevation	Miles from S. Francisco
Seminary Park	Cal		11	12.25	Towles	Cal		3704	150
Sentinel	Ariz		688	820.5	Tracy	Cal	200	64	71.73
Separ	N. M.		4503	1157.6	Traver	Cal	500	292	232.46
Sepulveda	Cal		461	473.5	Tremont	Cal		61	72.77
Sesma	Cal		227	212	Tres Pinos	Cal	500	514	101
Seape	Cal		450	476.6	Trigo	Cal			114.7
Seven Palms	Cal		584	590.9	Tropico	Cal		428	476.3
Shady Run	Cal		4160	164	Truckee	Cal	1500	5819	210
Shedds	Or	80		680	Tucson	Ariz	10000	2390	975.4
Shell Mound	Cal	40	10	7.96	Tulare	Cal	4000	282	251.12
Shellville	Cal	250		50.31	Tulasco	Nev	30	5484	607
Sheridan	Cal	150	115	127	Tule	Nev		4313	420
Shingle Springs	Cal			135	Tunis	N. M.		4422	1190.1
Shoshone	Nev		4634	487	Tunnel	Cal		1401	455 9
Short	Cal		459	488.6	Turlock	Cal	250	107	127 39
Slisby	Cal		112	169	Turner	Or	400		712
Sims	Cal	53	1387	311	Tustin	Cal	500	117	521.7
Siskiyou	Or		4130	414	Two Miles Sta.	Nev		4156	281
Sisson	Cal	250	3555	338	Tyler	Cal		244	276
Slatonia	Cal			302	Udell	Cal			73.02
Smithson	Cal		975	292	Ulin	Nev		5000	662
Snowdon	Cal		2675	382	Upper Soda				
Soto	Cal		186	200	Springs	Cal		2360	325
Soda Springs	Cal		6749	192	Vacaville	Cal	600		64.56
Sobrante	Cal		23	20.82	Vail	Ariz			997.8
Sobre Vista	Cal			57.21	Vallejo Junc.	Cal		12	29.01
Soledad	Cal	200	182	143	Valley Spring	Cal	60		131.32
Soquel	Cal	500	53	115	Vanarman	Ariz			1172.4
South Side	Cal		2350	431.9	Vanden	Cal		78	54 63
South Vallejo	Cal	5500	14	31.01	Vega	Cal			97
South Los Gullicos	Cal			64.51	Verdi	Nev		4495	234
Spadra	Cal	250	705	511.5	Vervain	Cal		165	496.8
Spences	Cal		79	125	Vina	Cal	150	213	205
Springfield	Or	200		646	Violet	Cal			66 45
Stanwix	Ariz		565	815.9	Vinvale	Cal		104	492 5
Stege	Cal		23	13.92	Vernalis	Cal			93.45
Steinman	Or		3035	421	Vincent	Cal		3211	421.5
Steins Pass	N. M.		4351	1119.4	Vista	Nev		4403	252
St. Helena	Cal	2500	254	64.36	Volcano Spr'gs	Cal		225	600 8
St. Joseph	Or			880	Wade	Cal		567	321.3
Stockton	Cal	20000	23	91.7	Wadsworth	Nev	661	4085	278
Stockyards	Cal		17	8.75	Walkers	Or			630
Stone House	Nev		4422	455	Wallace	Cal	150		121.32
Strauss	N. M.		4083	1272.2	Walters	Cal		195	625
Strongs Canon	Cal		6312	203	Walthall	Cal		65	100.7
Studebaker	Cal		102	497.2	Wanda	Cal		294	516.1
Suisun	Cal	1000	11	49.5	Wapato	Or			867
Summit	Cal	50	7017	195	Warfield	Cal			60.71
Sunol	Cal	200	264	36.60	Warm Springs	Cal		46	37 39
Surrey	Cal			104.17	Warren	Cal			376.1
Surbon	Utah		786		Watsonville	Cal	3000	23	101
Suver's	Or	25		916	Waverly	Cal		215	113.7
Sweet Water	Ariz		1996	902.1	Webster	Cal		26	5.44
Swingle	Cal			80.64	Wells	Nev	243	5628	614
Tacna	Ariz		325	770.5	Wells	Or	20		919
Tagus	Cal		292	246.55	West Berkeley	Cal	57c	14	10.42
Talbot	Cal		314	183.48	West Fork	Or			523
Talson	Cal			111.89	West Glendale	Cal			474.3
Tancred	Cal			101.67	West Oakland	Cal		12	5.89
Tamarack	Cal		6200	186	West San Leandro	Cal		20	15
Tangent	Or	70		685	West San Lorenzo	Cal		10	17
Talent	Or	25		437	Westley	Cal			101.25
Tartron	Cal			113.25	Wheatland	Cal	600	90	130
Taurusa	Cal			247.55	Whites	Or			887
Teal	Cal		10	44.19	White Plains	Nev		3894	314
Tecoma	Nev	60	4812	674	White Rock	Cal			119
Tehachapi	Cal	300	3964	361.7	White Water	Cal		1126	583.4
Tehama	Cal	500	222	187.34	Whitneys	Cal		139	114
Tehama	Cal	329	222	213	Whittier	Cal	500	239	503.1
Tejunga	Cal			466.5	Wildwood	Cal			63.81
T'mpleton	Cal	500	773	222	Wilbur	Or	100		583
Tennent	Cal		327	72	Wilcox	Ariz	500	4164	1065 3
Terrace	Utah	100	4548	709	Williams	Cal	450	84	125.19
Texas Hill	Ariz		353	793 2	Willows	Cal	1600	136	150.87
Thenard	Cal		31	503	Willsburg	Or	250		767
The Palms	Cal	100	140	494.6	Wilmington	Cal	950	9	504.5
Thermal	Cal			618.8	Wilmox	Ariz		2664	985.1
Tulshe	Nev		4170	283	Wilna	N. M.		4557	1170.2
Thompson	Cal			42 06	Winnemucca	Nev	2000	4333	414
Tipton	Cal	300	267	261 52	Winsted	Cal		5723	212
Toano	Nev	123	5975	650	Wluthrop	Cal		201	488.8
Tokay	Cal		267	256.75	Winters	Cal	500		76.82
Toltec	Ariz		1507	922 8	Wolf Creek	Or	15		504
Tormey	Cal		12	27	Woodbridge	Cal	250		106.31
Tortuga	Cal		183	677					

INDEX TO STATIONS—Continued.

STATIONS	State or Terr'ty	Population	Elevation	Miles from S. Francisco	STATIONS	State or Terr'ty	Population	Elevation	Miles from S. Francisco
Wolfskill	Cal			74.85	Yountville	Cal	500	107	55.05
Woodburn	Or	300		741	Yuba	Cal	600	71	141
Woodville	Or	100		465	Yuba Pass	Cal		5500	178
Woolsey	Nev		4008	348	Yuma	Ariz	1200	140	730.9
Woodstock	Or				Yulupa	Cal			57.91
Woodland	Cal	4500	63	85.95	Zayante	Cal		95	70
Wrights	Cal	150	990	82	Zuleka	Cal			399
Yolo	Cal	300	79	90.86	Zuni	N. M.		4187	1209.5
Yoncalla	Or	40		606					

COMPLETE INDEX TO STATIONS ON THE "NORTHERN DIVISION" OF THE SOUTHERN PACIFIC COMPANY.

CALIFORNIA.

Stations.	Population.	Elevation.	Dist. from San Francisco.	Stations.	Population.	Elevation.	Dist. from San Francisco.
Alamitos		191	58	Monterey	2500	5	125
Almaden	200	348	63	Monte Vista		740	119
Aptos	300	102	112	Morocojo		13	112
Baden		39	12	Mountain View	500	73	39
Bardins		48	115	Murphys		95	42
Belmont	250	31	25	Oak Grove		17	19
Bernal		188	4	Ocean View	500	293	7
Bolsa		177	89	Pajaro	100	23	99
Bradley	100	539	195	Paso Robles	600	723	216
Carnaden		158	183	Perrys		299	66
Castroville	600	7	110	Redwood	1800	9	28
Cholone		232	152	Salinas	2600	44	118
Chualar	100	103	129	San Andres		153	106
Coburns		259	158	Sanardo	150	452	182
Colma	100	171	9	San Bruno	50	15	14
Coopers		23	113	San Carlos		21	26
Coyote	225	251	63	San Francisco	350000	12	
Eden Vale		180	57	San Jose	25000	86	50
Fair Oaks		48	31	San Lucas	100	396	172
Gilroy	2200	193	80	San Mateo	950	22	21
Gonzales	350	127	135	San Miguel	700	616	207
Greystone		267	61	Santa Clara	3000	72	47
Hillsdale		147	55	Santa Cruz	7500	18	121
Hollister	2000	284	94	Sargents	100	135	87
Kings City	150	332	153	Soledad	300	192	143
Lawrences	150	64	44	Soquel	350	53	115
Loma Prieta	150	320	116	Spences		79	125
Madrone		342	69	Tennants		327	72
Martins		14	113	Templeton	500	773	222
Mayfield	900	26	35	Tres Pinos	200	514	101
Menlo Park	500	64	32	Watsonville	2500	23	101
Millbrae	200	8	17				

ADDENDA.

CONTAINING ADDITIONAL INFORMATION

RELATIVE TO

CITIES, TOWNS, AND POINTS OF INTEREST

DESCRIBED IN THIS VOLUME.

PASADENA.

Its Advantages over other Localities.   Its Climate and Growth. A Health Resort and Business Centre.

OUTSIDE of Los Angeles, Pasadena is more widely and favorably known than any city in Southern California. Other towns may possess many of its characteristics, but Pasadena claims to be the central point or focus of all that is good in the country. It is said, and truly, that one can find in New York almost everything obtainable in the entire country and a great part of Europe; so, as regards Pasadena, its citizens claim that it possesses the good qualities of almost every section of the state, without the bad ones.

The claims of Pasadena may be summed up briefly as follows: (1) A climate that challenges the world. If this may seem an exaggeration the reader is invited to communicate with the President of the Pasadena Board of Trade, who will provide the names of persons who have traveled over the entire globe and selected this section as the finest, all things considered. (2) Unrivaled scenic beauties. (3) Richness of soil and kindred advantages to the farmer. (4) A railroad centre. (5) The most fashionable winter resort on the Pacific slope. These are some of the features that have built up Pasadena from a village of 2,500 inhabitants four years ago to a busy city of 15,000 to day, and a winter population of from 18,000 to 25,000 souls. The people that come to Pasadena are of four classes. First, wealthy tourists; second, invalids, rich and poor; third, home seekers, who must earn a living, and fourth, wealthy home seekers.

The reader is presumably one of these. He or she is going the rounds of the Southern California towns in search of a home, health or pleasure, and wishes to know in a few words exactly what Pasadena has to offer. If you belong to the first class mentioned and have come to Pasadena to spend your time agreeably, this city has everything to offer. The Raymond Hotel is one of the best equipped west of New York, and boasts the finest scenery from its piazzas in the country. Painters', Websters', the Carleton, and several others, will give you the best of accommodations at different altitudes and prices. Four miles from the city the Sierra Madres rise to an elevation of from six to eight thousand feet. The range abounds in cañons and drives of the greatest beauty; falls, cascades, trout streams, caves, deep gulches, trails leading to the summit, and a thousand and one novel features, affording the tourist a new pleasure every day in the year. It is this feature which has made Pasadena the great fashionable winter resort of America. Tourists do not want climate alone; they are paying for amusement, and in its drives and innumerable natural beauties Pasadena is unexcelled. Nine miles from Los Angeles, twenty from Santa Monica and sea bathing, with the finest opera house in Southern California, the tourist has every facility for pleasure and enjoyment. The hunting is good, and out door life can be counted on nearly any day in the true land of flowers.

What can the invalid expect? Pasadena lies about twenty-five miles from the ocean, at the head of the San Gabriel valley, environed by mountains on three sides. The altitude of South Pasadena is about 800 feet; of North Pasadena, at Wilson's Peak, about 6,000. Good hotels and boarding houses are found up to the

2,500 feet elevation, with camps higher up ; so you can take your choice. Pasadena winters, from November to May, remind one of New England or Ohio in October. It is cold often in the morning and evening, and you will sometimes see frost and a little ice in low places ; but roses and all the plants bloom every day in winter, which tells the story of the climate. It may seem cold, but when flowers grow out of doors, and semi-tropic ones at that, the invalid need not fear a blizzard. The mean for winter, taking January as an example, is 53 ; that of Jacksonville, Fla., 55. There is rarely a wind sufficient to blow off one's hat. One thunder storm a year is the average, and then in the mountains only, as a rule. The sudden changes of the east are unknown. The difference between winter and summer is expressed by fifteen degrees. The seasons melt one into the other. There is an almost entire absence of sudden storms. Cyclones, blizzards, tornadoes, as they occur at the east, are here unknown.

The winter is the rainy season, during which twenty inches of rain falls here upon an average. The annual fall at Jacksonville, Fla., is 55.94. Hence there is no malaria-producing element here. The climate is moderately dry—the mean annual humidity is 56. The winter begins with the vintage, the country grows green, and ends with the flowering of the peach and other fruits. At Christmas the wild flowers are at their best and the orange groves burdened with fruit.

·The Pasadena summer is far more comfortable than the same season in any eastern city. The mean summer temperature, taking July as an example is 67; that of Jacksonville, Fla., 83. Sunstroke is unknown. The nights are invariably cool. A casual glance at these features will show that Pasadena claims the good features of all other health resorts, except those, as Yuma, where the maximum of dryness is found. Jacksonville, Fla., offers a season of from November to April with semi-tropical conditions and possible malaria. Invalids must, as a rule, go north in summer, so cannot spend any extended season in the spot of their choice ; yet it is well known that consumption cannot be cured in six months.

Pasadena, on the other hand, offers a residence winter and summer under the most favorable conditions, and more, in a locality where the poor man has a chance of being self supporting in a rich and rapidly growing community. It offers the widest range of climates. In two hours from Pasadena one may find the sea shore, or localities six or eight thousand feet above it ; and almost every possible condition, from the elevated, hot *mesa* to the deep, moist cañon. It is this variety of climates which has given Pasadena its world-wide fame. Pasadena, then, is the land of out door life. Here one finds at least three hundred and thirty days in the year which the invalid can spend in the open sunlight, and thus receive the only true cure for lung troubles.

The greatest test of the curative property of Pasadena's offerings lies in the number of persons who have been benefitted by coming here. They are legion. She does not offer a panacea for all the ills that flesh is heir to ; or claim to restore invalids given up by doctors in the east ; but to those who come in time and take the advice of competent local physicians, every inducement is offered.

HOME SEEKERS.

The third class interested in Pasadena are home seekers who are dependent upon their labor for support. These are pouring into Southern California in a steady stream, and Pasadena itself is one of the results. The workingman will find here a city of 15,000 inhabitants, swelling in the winter season to almost twice the number. The city, formerly a vast orange grove, is now laid out into streets and blocks, with miles of cement sidewalks and graded streets, four or five lines of horse cars, electric lights, elegant churches of every denomination, secret societies, five or six banks, four distinct railroads—two being transcontinental, opera house, hotels, club houses, rows of brick blocks, elegant villas, manufactories and industries of all kinds. In fact, Pasadena lacks but one thing, which you will find in every perfect town in the east, and this is the saloon. It is a temperance town. There are schools, young men's Christian association, and half a million dollars invested in churches and church property, but not one cent in the liquor business.

Without going into particulars, almost every industry found in eastern towns obtains here; and the workingman, the professional, artisan, or other worker, will find the same opportunity here as elsewhere. The city is barely two years old in its present form, yet it has brick yards, planing mills, fruit canning establishments, steam laundries, gas works, cold storage warehouses, ice manufactories, and business enterprises of great variety, representing a vast amount of capital.

The workingman, perhaps driven from the east by severe winters, where the cold is a menace, as winter comes on finds here a great contrast. If he cannot afford a fire his children feel chilly on winter nights instead of freezing, and summer comes every winter day from 9 a. m. to 4 p. m. The farmer or agriculturist finds an open season the year round. While in January in Ohio he was snowed up, Christmas here finds barley either up or ploughing going on. There is something growing all the time; six crops of fodder (alfalfa) for the cattle, and other things in proportion. In Florida, grass, milk, apples, pears, nuts, butter, peaches, etc., are scarce or unknown. In a Pasadena workingman's home of from one to ten acres, you will find the following, and note carefully the contrast, as it tells the story in a word of the agricultural possibilities of this section: Apples, guavas, peaches, grapes, oranges, currants, limes, strawberries, and all small fruits, loquat, pomegranates, pears, walnuts, chestnuts, bananas, almonds, lemons, figs, and every fruit, flower, shrub, or tree, from the semi-tropics to the shores of the northern ocean. In fact, fruits and flowers of all kinds find common ground here; the cork tree and the sturdy pine grow side by side with the magnolia and banana.

The orange, lemon and grape industries are the most important and constitute the out door industries in which the most capital is invested. Near Pasadena are two of the largest wineries in the country, while sheep, horses, pigs and ostriches are among the valuable live stock. The settler will do well to investigate Pasadena and its outlying country before selecting a home. Its schools, society, and lack of saloons commend it to every thinking man and woman.

WEALTHY HOME SEEKERS.

Pasadena is one of the wealthiest places in Southern California. Its beauty, its grand position, environed by hills and mountains, with views or vistas unparalleled stretching away, have attracted wealthy and cultivated people from all over the world. who have selected homes here and erected elegant villas, costing from ten to fifty thousand dollars. Pasadena boasts that over twenty millionaires spend part of their money here, and in all probability no place of its size west of Chicago possesses so many wealthy men. The city is made up in a great part of the homes of wealthy men. Orange Grove avenue, Colorado, and the adjacent streets and avenues show what taste, culture and unlimited wealth can produce. The social conditions are as perfect in Pasadena as can be found in any city of the east a century old. There is no rough element here. Many judge this section by Colorado and other western towns, but the opposite is the case. Pasadena, in its social organization, is made up of the cream of other cities of the Union, and so presents an attractive outlook to the home seeker. Further inquiries regarding Pasadena may be addressed to the President of the Pasadena Board of Trade, Pasadena, Los Angeles Co., Southern California.

POMONA, SOUTHERN CALIFORNIA.

Midway between the cities of Los Angeles and San Bernardino, encompassed by picturesque foot-hills, over which the Cucamonga mountains, San Antonio peak, Old Grayback and San Jacinto, like sentinels, stand guard, is the beautiful and fertile section of Pomona. At a point where a low spur of the coast range, called *San Jose hills*, debouches into the valley, penetrating almost to its center, the *City of Pomona* lies at our feet, with its fine broad avenues, lined and shaded with the majestic evergreen Eucalyptus and the beautiful fragrant pepper trees, the semi-tropical palms and magnolias, its many miles of cement walks, beautiful homes and cottages, nice flower and grass plots, encircled by its grand and beautiful *orange and lemon groves* and thousands of acres in peaches, pears, prunes, apricots, nectarines, plums, pomegranates and all varieties of other fruits and berries. For the lover of horticulture and fine homes a grander sight can not be seen.

Pomona contains an intelligent population of *five thousand* refined people. Its church and school privileges are excellent. There are eleven church buildings, each supported by large flourishing congregations.

The Climate has few equals but no superior in Southern California. The mountains to the north and northwest shut out the cold and hot desert winds entirely. The breeze from the ocean—which we have daily from May to November—thirty-five miles away, is less moist than near the coast, and is subdued and softened by sweeping over the intervening low hills and warm plains.

Temperature. The mean average heat of July and cold of January, in the principal cities of Southern California, is as follows:

	Cold.	Heat.	Difference.
POMONA,	52°	68°	16°
LOS ANGELES,	52	67	15
SAN DIEGO,	52	66	14
SANTA BARBARA,	52	66	14
SAN BERNARDINO,	51	70	19

The daily mean temperature at Pomona for May, June, August and September is respectively 62°, 64°, 68°, 72°, testifying that days of extreme heat seldom occur.

Come to Southern California over the "Santa Fe" or the "Sunset" route, get a stop-over at Pomona, and see our *Loveliest Valley of the Plains*.

Educational. The *Pomona College*, now being erected will be an institution of learning second to none on this coast. A brick and stone building of architectural beauty, 103 x 81 feet three stories and basement. The basement was completed (Sept. '88) and work on the building will progress rapidly. The trustees will erect buildings of equal character for the various departments as needed. The brick for the building are made on the grounds, and the stone quarried near by. The college is located on *Piedmont Mesa*, overlooking the whole valley. Altitude 1500 feet above the sea. The faculty has procured temporary quarters in the city and the collegiate course is now taught by an able corps of professors.

Our *Public Schools* are graded. There are four large school buildings, and the plans are made for a fine large brick school to cost $35,000. The attendance is over six hundred pupils, who are taught by an able corps of twelve teachers.

Water. There are three sources of supply for irrigation, viz.: mountain creek, springs and artesian wells.

The water from the mountains—San Antonio Creek—piped a distance of ten miles into the valley—a system complete in itself—is a large source of supply. Numerous springs encircle the valley fed by subterranean streams from the high mountains. These sources of supply are yet to be fully developed, as no occasion has arisen to demand it. There are in this valley, within two miles of the City of Pomona, over *one hundred artesian wells*, giving an undiminished flow.

Our *Pipe System* for distributing the waters *is the most complete in Southern California ;* there are now over *seventy-five miles* (396,000 feet) *of pipes laid*—no open ditches—to convey these waters to the highest point on each 10, 20 or 40 acre tract of land, without loss of leakage or evaporation, and new lines of pipe are constantly added as needed.

The perpetual right to use these waters—free of cost—for irrigation is sold with the land.

The abundant supply of water and the excellent distribution of the same, surpasses any other settlement or system in Southern California.

Fruits. The fruits of Southern California are known world-wide. The combination of rich, mellow soil, well watered, with sunny, balmy atmosphere, can produce nothing less than the most delightful, luscious fruits of all kinds ; and some—notably the citrus and semi-tropical—of superior excellence. All deciduous fruits are, and have been grown here with the greatest success. Berries and small fruits ripen early. Orange and lemon trees grow luxuriantly, and the fruit commands as high prices as any in the market. The upper lands are especially adapted to the culture of these fruits as well as the fig, olive, and all semi-tropical fruits. The enormous profits of olive culture are almost incredible, and invite the general cultivation of this beautiful tree and profitable fruit in this locality. Besides the orange, lemon, olive, fig berries and other fruits, we raise as fine vegetables, corn, alfalfa, etc., as the world can produce.

Fertility and Productiveness of the Soil. The soil is generally of sandy, gravelly loam, very deep, easily cleared and cultivated, and very productive. From its composition and admirable drainage no danger of malarial diseases is to be feared from irrigation as in soils of a heavier texture or adobe formation. Owing to this peculiar formation tourists and others find driving upon the thoroughfares a source of pleasure, because of the absence of dust in summer and mud in winter.

Lands with free water-right are worth $150 and upwards per acre; improved, $250 to $1,500—according to locality and improvements. Moist lands in the lower valley suitable for deciduous fruits and vegetables without irrigation can be had for $125 and upwards per acre.

An authority says : "We believe—no panic or calamity interposing—that in the next ten years it will be very difficult to buy any desirable lands with water, suitable for orange or raisin growing, in Southern California for less than $1,000 per acre."

Railroads. Pomona is destined to be a great railroad center of no mean importance. Already we have the two great trans-continental systems, the " Santa Fe" and the "Sunset" routes, running through our valley and city, both having fine large depot buildings. All trains on these routes make regular stops. This is the terminus of the Pomona, Elsinore and San Diego R. R., now partly graded, on which cars will be running within six months. This, being an independent line, will have machine shops and terminal facilities located here. Pomona is, also, the terminus of the Pomona, Olinda and Anaheim R. R., giving us four independent lines of road direct to the grand Pacific ocean. We have a steam motor line with three miles of road in operation and will soon be extended to the mountains, also four independent horse street car lines with over nine miles of road in operation, diverging in separate directions, on which cars make frequent regular trips every day of the year.

Business. We have eight large hotels and lodging houses with ample accommodations for the tourist and traveler. The Pomona Fruit Co. have erected a two story brick building, 100 x 32 ft., with basement and additions ; have a capital of $50,000 ; employ 100 to 200 hands—many of them ladies—(no Chinamen), and put up twenty tons of fruit daily. They have bought additional grounds and will erect more buildings this winter and add mills for the pressure of olive oil. We have three banks, a large opera house, public hall, secret society halls, sash and door factory and planing mill, two iron pipe factories, cement pipe works, steam laundry, two wineries, brickyards; granite, lime, sand and brown stone for building purposes near the city ; large fine stores, well filled with goods ; all kinds of trades people; streets and buildings lighted with gas.

Mountain Scenery. The scenery is grand as well as beautiful—beauty and grandeur, as it were, combined in one sweep of the vision. The lofty Sierra Madre range north presents strikingly sublime scenery while the immense plain stretching away to the southward, diversified by rolling grass-covered hills and lesser mountain ranges, orange groves, orchards, vineyards and attractive homes, around which perpetual flowers bloom, is a prospect which gives a thrill of new life to the invalid, starts the sluggish current onward with fresh vigor, and paints a flush of returning health on the faded cheek. The day is not far hence when this eminently suited locality will be utilized for the benefit of the thousands who will come here seeking health. Those who have a tendency to pneumonia, bronchial, catarrhal, or asthmatic affections, or those otherwise in delicate health, would do well to avail themselves of this great natural sanitarium, whose atmosphere is as pure as the breath of innocence and whose zephyrs bear healing upon their wings.

LAMANDA.

A Delightful Resort.

The Gem of the San Gabriel Valley.

Lamanda is situated 12 miles from Los Angeles, on the Mesa, north of San Gabriel proper, and east and adjoining Pasadena in the San Gabriel Valley. Its elevation varies from six to eighteen hundred feet, and it contains an area of over 6,000 acres. In all Southern California no other spot is so delightful as the San Gabriel Valley, and to no other spot do a tithe of the tourists resort that cluster here. It was in this valley that flourished the most prosperous of all the many missions, founded by the Franciscan friars over a century ago. Of all this fair region they chose this valley as the most desirable, and although the land remained in the possession of the unprogressive Mexican for so long a period, time has proved to the world the wisdom of the choice made by these priests of the olden time. This valley is justly called the "Italy of America"

The valley proper extends east and west for twenty five miles on both sides of the San Gabriel River, and from the Sierra Madre Mountains to the ocean; but it is that portion lying south of the mountains, and comprising a strip of land fifteen to twenty miles wide and forty miles long, that constitutes the "Garden of Eden" of modern times.

Here grow, side by side, the Norway pine and the banana, the camphor tree and the apple, the elm and the palm. The perfection of flowers and shrubbery, the beautiful lawns and gardens, where almost every variety of trees, plants, and flowers in the world may be found, could have been grown in no other land in so short a time, if at all. From autumn to spring and from spring to autumn, there is no cessation in the growth of vegetation. There are no frosts to blast, no winters to destroy even the most delicate plants.

Everywhere are cypress hedges, the tall eucalyptus and the spreading pepper trees, acacias and grevillas, giant palms and cacti, rose trees and calla lilies, marguerites and magnolias, with trees, shrubs and flowers of every description, from Australia and New England, from every tropical and every temperate clime.

To the north rises the Sierra Madre Range, its summits reaching up among the clouds to a height of 7,000 feet. The scarred and seamed outlines of these mighty monuments, that guard the valley from arctic evils, present a picture of awe-inspiring grandeur and sublimity—a picture that is unsurpassed even among the Alps. Down the cañon-creased sides of these eternal hills ripple the cool mountain streams, now laughing along in foaming cascades and anon wreathing some precipice with rainbow spray as the sparkling waters take their wingless leap down to the ragged rocks a hundred feet below, and then murmur adown the widening cañon and under the spreading branches of gnarled and picturesque live-oaks, which seem as aged as the gray granite boulders whose fantastic figures frown around. Scenes of rugged beauty and pastoral enchantment everywhere alternate. No other land is so lovely as this valley; no other spot knows such ideal, happy homes.

The northern portion of this Eden-land has an altitude of from 1,000 to 1,500 feet above the level of the sea, and the atmosphere is exceedingly pure and entirely free from malaria. Its dryness renders the air especially beneficial to those whose lungs are diseased. Invalids come here by the hundreds, and in every instance, where they are not past all hope, they speedily find that precious boon which they have sought in vain in every other clime. Remarkable, indeed, is the record of cures wrought by this wonderful climate. Consumptives, whom physicians of the East had declared past all help, have come here, and in a few weeks have shaken off the fetters of that Eastern ice-born curse, and are to-day enjoying perfect health. Is it strange that they are happy? that they love this sunny southland? Would to God that the hundreds of thousands in the East who are slowly dying might come hither! What is more blessed than to see the light of hope wake in the invalid's eye, and the flush of returning health spread over the cheek, to note the form grow supple and the step elastic, to watch the smile of happiness and contentment grow over the careworn visage with the assurance of complete restoration! The San Gabriel Valley is an Eden to him who possesses health, a Paradise to him who here finds it. The soil is remarkably well adapted to fruit culture, and it contains some of the largest fruit ranches in Southern California. The water system is one of the best, the supply coming from

Precipice Cañon, a never-failing mountain stream, and is piped to nearly all parts of the town. Good wells can also be obtained at from 60 to 100 feet in some portions. Of the immense immigration that has come to California in the last few years, Lamanda has received but a small proportion in numbers, but a much larger proportion in wealth than many other sections of the San Gabriel Valley, the result of which is a number of elegant villa residences that would do credit to any country, noticeable upon the higher elevation where the grand surrounding scenery and entrancing view make property for residence very desirable. The railroad service from Lamanda is excellent. Situated, as it is, on the main line of the Santa Fé, and also the terminus of the Pasadena Branch of that system, give it the benefit of all trains over these lines. There are now eight trains a day each way, and a theatre train three nights a week.

South of the railroad lies the celebrated Sunny Slope Estate, the late home of the Hon. L. J. Rose, now the L. J. Rose Company (limited), an English company with a capitalization of £1,000,000. This ranch contains 650 acres of grapes, 150 acres of orange grove, besides other fruit, grain fields, etc.; and produces annually over 300,000 gallons of wine and 100,000 gallons of brandy; its largest orange crop amounted to about 100 car loads of 30,000 boxes; its other produce, such as hay, grain, etc., is for use upon the ranch.

The machinery and apparatus connected with the winery and distillery are all of the most improved patterns. The grape crushers have a capacity of 100 tons of grapes per day, of which there are two. The quality of wine and oranges is well known throughout the United States, and is destined under its present management to soon become world renowned.

As a place of residence Lamanda is unexcelled. The climate, while it is everywhere delightful, varies in different localities. The severe hot weather which is supposed to exist here is much of a myth, save in those valleys which lie east of the first range of mountains where the mercury sometimes registers more than one hundred, though rarely; while even that far inland something of the influence of the ocean breeze penetrates, which, with the altitude, renders the nights exceedingly pleasant. In the valleys that extend upward from the ocean toward the mountains, and at a distance of from fifteen to fifty miles from the ocean, is to be found the perfection of climate that has rendered this region famous throughout the world. There are three railroad stations in Lamanda, viz.: Marceline, Fair Oaks, and Lamanda. The latter is the principal one, and situated at the junction of the main and branch line of the Santa Fé Railroad. It is called "Park" from the stately oak that abound in the neighborhood. Here are located the postoffice, stores, stables, hotel, etc. The stages from the Sierra Madre Villa, and other hotels, connect with the trains at this station.

The Sierra Madre Vintage Company's Winery is established here. It has a capacity for crushing fifty tons of grapes per day, and manufacturing a superior quality of wine and brandy.

The cosy, home-like Brightwood Hotel is directly opposite this station, and from its central position and excellent table has established a growing reputation. The home-like character of this ho-telry makes it an especially charming place for tourists' headquarters. From the shady verandahs of this hotel the tourist can see a panorama of most beautiful and attractive scenery. Directly to the north are the Sierra Madre Mountains, to the east stretches the lovely San Gabriel Valley, dotted here and there with pretty villages, whose church spires gleam whitely against the blue sky, towering upward from the dark green masses of goldenfruited orange trees. The hotel is equally distant from Pasadena, Old Mission, Sunny Slope vineyards, Chapman's orange orchards, Baldwin's great ranche, Sierra Madre, and the romantic cañons of the Sierra Madre Mountains. All of these places are within an easy carriage ride, and if one have but a day or two to spend in this valley the "Brightwood" is the best place to make one's headquarters, both for convenience and comfort. Economy, both of time and money, can be subserved by directing one's excursions to points of interest in the San Gabriel Valley from the Brightwood. Near the hotel is the stables of Henry Eaton & Sons, where good carriages can be obtained at reasonable rates, and, if desired, drivers well acquainted with all places of interest in the valley.

Fair Oaks is on the Pasadena branch of the Santa Fé, and the nearest station to "Fair Oaks," the home of Hon. J. F. Crank. This grand ranch contains five hundred acres, and the homestead of Mr. Crank is a bewildering bower of beauty.

Marceline is on the main line of the Santa Fé. It is owned by several wealthy gentlemen, but has not as yet been put on the market. It is an oak grove, containing about 250 acres, and will at no distant day be the site of many a fine residence.

MONROVIA.

The Home of Health and Pleasure.

In the Heart of the Italy of America.

Monrovia is centrally located in the San Gabriel Valley, and is unquestionably a city of great attractions as is to be found in the fairest valley on the coast. It is situated at the base of the Sierre Madre Mountains, on a gentle elevation, and commands a view of the valley for miles in either direction and of not less than a dozen villages. The mountain view has often been pronounced the finest in the State, and the ocean that is visible away to the south, through a break in the Puenta Hills, sends its cooling breezes to fan the valley into refreshing healthfulness. It is situated eight miles east of Pasadena and seventeen miles northeast of Los Angeles on the through line of the California Central, the Atchison, Topeka & Santa Fé route from Kansas City to Los Angeles, and on the surveyed line of the Southern Pacific Railroad. The site of the town is less than a mile from the base of the Sierras, and the land slopes gently away to the east, west, and south. From its high elevation (1,200 feet above the sea) a magnificent view is had of the valley below, that slopes in a southwesterly direction to the ocean, distant about thirty miles. To the right lies "lovely Pasadena," with its fine buildings, the great Raymond Hotel crowning an isolated hilltop like some ancient castle; still nearer is the long famous Sierra Madre Villa, with its elegant grounds, and the villages of Alhambra, Lamanda Park, Sierra Madre and Arcadia. To the left are Duarte and Azusa, separated by the San Gabriel River, whose course is traced down through the valley by the shimmering white sands. Farther off are Los Angeles, the "City of the Angels," and the numerous surrounding villages. Still farther on, Santa Monica, San Pedro, Wilmington, and Long Beach nestle by the side of the restless mighty ocean, out on whose bosom rise the blue outlines of Santa Catalina and San Clemente Islands.

The soil is sandy loam, and peculiarly well adapted to raising both citrus and deciduous fruits, which grow to perfection in this vicinity. Monrovia is but little more than three years old, the first lot on the site of the town having been bought on the 17th day of May, 1886. The growth of the place has been phenomenal, it now having a population of over 2,000. There are two street car lines, an electric road, and the right of way has been secured for two other roads A large and handsome school building, two fine churches—the Methodist and Baptist, costing $8,000 each; the Grand View Hotel, one of the finest and best conducted in the valley; several other hotels, two handsome banking buildings, the Granite Bank being one of the finest in the State. The location of Monrovia has been noticed. The high and rolling ground gives it a much more sightly view than can be obtained from any of the surrounding towns. The ocean and mountain breezes both prevail, and give the place a remarkably pure and wholesome atmosphere. It is a generally admitted fact that the highest elevations in the valley are the most healthful, and, as Monrovia is among the valley towns of greatest elevation, being several hundred feet above Los Angeles or Pasadena, it is preferred above most other places by invalids. Here one may spend months in genuine comfort in the enjoyment of the picturesque scenery of the valley and the rugged grandeur of the mountains. If he catches the spirit of California enterprise, which all residents have and all tourists get, and invests in some of the desirable surrounding property, he will become richer in wealth, as he certainly will in health, with each additional month spent here.

SANTA MONICA.
The
"Gem City by the Sea."
A Charming
Watering Place.

Santa Monica is situated directly on the shore of the Pacific Ocean, distant seventeen miles nearly due west from Los Angeles, and about four miles north from Port Ballona. It contains an area of nearly five square miles, having a beach frontage of about two and a half miles, by two miles inland. Its population at present is estimated at 1,500, with a transient population of about the same number, making in all an average population of 3,000.

The superior climatic conditions of Southern California, as a whole, is a feature which, as it becomes better known, is commented upon and admired by people from all parts of the world. Southern California's great prosperity at present is due largely to this fact. It is not enough to say that competition in fares and freight to this coast has brought this wonderful development; while this has certainly been a large factor, still it is not the sole cause. Had Southern California not possessed superior natural advantages for residence and business purposes, the immense number of people from all parts of the East who came here during the recent low rates of fare would not have remained and invested their means.

The townsite of Santa Monica comprises a part of the famous rancho San Vicente, a large holding of 30,000 acres or more. In 1875 the Hon. John P. Jones, the famous Nevada mining king (and who has since become United States Senator from his State, who at that time owned large mining interests in the Cerro Gordo district, situated near the Nevada line of California, two hundred miles distant from the coast), being desirous of an outlet by rail from his extensive mines, determined to build a railroad to Los Angeles and the coast. Accordingly, he came down, "looked the field over," and purchased a three-quarters interest in the famous ranches known as the "San Vicente" and the "Boca de Santa Monica," two large Spanish grants, comprising 36,000 acres of land. The price paid was $155,000, which amount was to be expended toward building the above-mentioned road, since known as the Los Angeles & Independence road. Work on the road began immediately, and was pushed rapidly.

A small tract, comprising the original townsite of Santa Monica, was subdivided into lots of 50x150 feet, and sold at auction, the sale beginning on the 15th day of July, 1875. The first lot sold was purchased by E. R. Zamoyski, for which he paid $510. The auction sale, which continued for three days and nights, took place on the grounds during the day, and at the Pico House, in Los Angeles, at night. During this time about twenty blocks of lots were sold, ranging in price from $125 to $510, aggregating nearly $200,000. These lots have since become the principal business properties of the town, and have increased in value more than ten-fold.

Things went well for a time; work on the road progressed rapidly. A wharf 1,700 feet in length was built in the meantime, capable of accommodating the largest of Panama vessels, many of which lay there for days and weeks discharging their cargoes. The wharf being built and the road completed as far as Los Angeles, the first train passed over the road from Santa Monica to Los Angeles in September, 1875. This marked an epoch in the history of Santa Monica long to be remembered.

Santa Monica's prosperity for the first two years of its history was, indeed, phenomenal. The freight traffic was large, passenger movements were heavy, real estate sales were rapid. The building operations were active; houses, both for business purposes and residences, sprang up on every hand, as if by magic; land enhanced in value rapidly; industries of almost every kind were established, and withal Santa Monica enjoyed a period of prosperity, the degree of which is seldom attained—even by new towns of Southern California to-day. Had the circumstances surrounding Santa Monica been favorable rather than otherwise, had the wharf and railroad continued under the management of their projectors, thus securing to the new port the bulk of San Francisco's steamer shipments to Southern California, as well as the foreign ocean passenger and freight traffic, the present commercial importance of Santa Monica can only be surmised. It is safe to say that it would have been one of the largest cities of Southern California.

While Santa Monica's climate has many features common to all of Southern California, still some features are peculiarly its own. The temperature is very much less variable here during the day, and during the season as well, than at most of the interior towns. Thus the thermometer rarely indicates a temperature below 40 degrees above zero, even in winter, and seldom reaches 90 degrees in the shade, during the warmest days of summer. At no time during the season does the temperature vary throughout the day more than 20 degrees. A constant sea-breeze fans the shore for miles distant the year round. Thus one continuous summer is realized in this favored spot, where flowers bloom constantly and fruits ripen in every month during the year.

The soil of this valley is wondrously fertile. A plentiful supply of water abounds throughout the valley, products of all kinds common to semi-tropical climates grow luxuriantly and yield largely. The soil near the coast is of a rich, sandy loam; as the mountains are reached, it partakes slightly of the nature of adobe, a very rich, dark, clayey soil.

The water supply of Santa Monica is equal to that of any town or valley in the State. There is enough water in the valley, if developed, to irrigate from 5,000 to 10,000 acres of land, besides supplying for domestic purposes a city of 50,000 inhabitants. The source is the streams from the Santa Monica Mountains, two miles distant from the town, nearly due north. Three to four streams are found in these mountains, which by development will yield 500 miner's inches, equal to a constant flow of 6 500,000 gallons every twenty-four hours. The source being situated 300 feet above the town site, a pressure of 150 feet per mile is obtained sufficient to throw a stream to the top of the highest building. At present the water is conveyed through iron pipes along the principal streets of the town.

The railroad facilities of Santa Monica are excellent. Being on a branch line of the great Southern Pacific trans-continental system, it has direct communication with Los Angeles and other commercial cities.

Through the kindness of the courteous and genial station agent, the following facts were learned: Number of trains arriving and departing daily, eight (three passenger and one freight train each way). On Sundays twelve passenger trains arrive and depart regularly, to accommodate the immense passenger movement during the watering season. To accommodate the rapidly increasing demand for more yard-room the company will soon have completed two miles of side track.

The agricultural and horticultural products of the valley surrounding Santa Monica are all that could be desired. Fruits of all kinds, both citrus and deciduous, including oranges, lemons, limes, bananas, figs, peaches, pears, apricots, nectarines, quinces, etc., grow luxuriantly and yield largely. Small fruits of all kinds, including strawberries, blackberries, raspberries, currants, gooseberries, etc., are grown extensively, and yield handsome profits.

Vegetables of all kinds, potatoes, cabbage, turnips, water melons, pumpkins, beets, lettuce, asparagus, etc., are grown largely and yield bountifully.

Grain of all kinds, including wheat, barley, oats, rye, corn, etc., also alfalfa hay, yield largely.

The superior advantages of Santa Monica for residence and business purposes, are readily apparent. Its unsurpassed climate, magnificent water supply, proximity to Los Angeles, superb mountain scenery, excellent surf-bathing, and great diversity of agricultural and horticultural products, all combine to make this one of the most desirable places for homes in Southern California.

⇘·Long ✢ Beach⇙

Long Beach is twenty two miles south of Los Angeles, on the S. P. R. R., situated upon a bluff of medium altitude, overlooking San Pedro Bay and the Ocean toward the south, with Santa Catalina Islands in plain view twenty-five miles out to sea.

To the east, north and west is a magnificent view of mountain scenery, Santa Aña, Sierra Madre and Santa Monica Ranges, which bound the great valley on the three sides.

Long Beach is the chosen sea-side resort for a number of prosperous cities and towns, Los Angeles, Pasadena, Monrovia, Riverside, San Bernardino, Ontario, Pomona, Whittier, Santa Aña, Orange, Anaheim and others.

The Beach at low tide is hard, smoothe and level, making the grandest boulevard on earth.

For a distance of seven or eight miles, twenty teams can drive abreast, the sand being so firmly packed by the action of the tide, that the wheels of the carriages make little or no impression upon it. Not unfrequently the shore bordering the water's edge is strewn with millions of little clams, the shells of which are extremely handsome; no two exactly alike in marks, color or shading.

Other varieties of pretty shells are also to be found, the searching for which is a pleasing and exciting pastime.

Large schools of porpoises and sea lions are frequently to be seen sporting on the water near the shore.

Long Beach has an intelligent, refined and moral class of citizens, excellent public schools; three church societies. *No saloons;* enterprising business men and a live newspaper, "THE LONG BEACH JOURNAL."

"AS OTHERS SEE US."

"The beach is positively the finest in the world; I have dipped in the water at Trouville, at Brighton, have dived in the surf at Long Branch and Coney Island, and of course have not neglected Monterey or Santa Cruz, but there is no beach like Long Beach, and this I claim will be conceded by every fair minded person.

"Standing upon the shore looking in a south easterly direction as far as the eye can discern, lies the broad expanse and boundless deep of the Pacific Ocean, heaving and swelling with majestic pride, as it bears upon its indigo-tinted surface innumerable ships, ocean steamers and seafaring crafts of every variety to and from all parts of the globe.

"Add to this scene the commanding and stately outlines of the Catalina Islands, anchored by nature in the depth of the sea, twenty miles distant, and the snow-white surf rushing headlong upon the beach, wafting the briny-tainted waves upon the numerous bathers arrayed in their variegated costumes, you behold a scene which cannot be portrayed by the artist's touch or possibly conceived by visiting the sea shore at any other point yet discovered. Such is the verdict of the thousands of people who enjoy the pleasures of Long Beach annually, many of whom have visited every sea-side resort of note upon the continent."—*Editor of Pasadena Call.*

"I think Long Beach has the best sea beach I have ever seen. It is certainly better in every respect than Coney Island, Far Rockaway, or Cape May. It is immensely superior to Nantasket. It is more attractive than Newport. One need not go to Birkenhead, or Deppe either. They can be found in California, if you will look for them."—*Prof. J. W. Redway, Geographer of New York.*

"All in all, taking Long Beach city and Long Beach sea shore resort as a whole, it is our opinion that no more favored spot can be found on earth, affording as it does a combination of unequaled climate, mountain, valley and ocean scenery, surf bathing, etc."

"We never supposed it possible to visit a point where the beauties and natural advantages surrounding would so completely overwhelm us with admiration. We could write a book in expatiating upon this lovely spot, and then the half would not be told."

ONTARIO.

The Foot-Hill Paradise of Southern California.

ONTARIO is situated in the County of San Bernardino, on the southern slope of the Sierra Madré, just on the borders of Los Angeles County, and enjoys an altitude varying all the way from 900 to 2500 feet above the level of the sea. It is on the height of land between the San Bernardino Range of Mountains, 40 miles to the east, and the Pacific Ocean, 40 miles to the west. It occupies this entire ridge, extending from the Sierra Madré eight miles south, with a width varying from three to four miles, and overlooks the great San Bernardino Valley in all directions, affording a most entrancing view of mountain, valley, foot-hill and plain, with towns and orange groves everywhere intermingled. The four highest peaks of Southern California are always before the eye, viz: Mount San Antonio (Old Baldy) adjoining the Ontario Tract on the north ; old Grayback and Mount San Bernardino to the east; and San Jacinto to the south-east. These majestic snow-capped peaks, towering above their fellows and glistening in the brilliant sunlight, afford a pleasing contrast to the luxuriant semi-tropical growth of the cultivated valley. Here on the upper slope of Ontario we find Orange Groves breathing their delightful fragrance upon the balmy air, free from any suggestion of cold and beyond the reach of blighting frosts, whilst eight or nine miles away the eternal snows keep their silent vigil. Nowhere, probably, on the face of the Globe are Winter and Summer brought into such close juxtaposition. Perpetual summer and eternal winter clasping hands across San Antonio Cañon ! The spectacle is an extraordinary one, and the more it is considered the more wonderful it becomes.

THE BEGINNING.

Six years ago the 17th of March of this year the initial improvement was begun in the now well-known and justly celebrated " Model Colony" of California, by laying the corner stone of the first building. Then there was nothing visible but the bare plain, hemmed in on every side save the west by the everlasting mountains, which afford such effectual protection from the rude northern and eastern blasts ; the openings through the modest Coast Foot Hills inviting the soft and balmy Pacific breezes—the veritable " Winds of the Western Sea"—into this mountain embraced Garden of the Gods. But six short years have witnessed a marvelous transformation indeed. The uninhabited wilderness, treeless and desolate, with no house anywhere within the range of vision, the playground of the Jack Rabbit and the home of the Coyote, has been, in all truth and soberness, made to blossom as the Rose. The wand of the magician, water, has been waved over the land. Industrious, thrifty settlers have been charmed by its scenic grandeur; captivated by its wondrously perfect climate; impressed by the abundance of its pure, clear, sparkling water from the cool grottos and crystal mountain streams ; and altogether won by the extraordinary wealth lying latent in its deep, rich orange soil. So, to-day, instead of a dreary waste of sage brush, without a solitary habitation to break the monotony, are now to be found two flourishing towns, with palatial brick blocks, schools, colleges and churches ; and pleasant homes bedecked with

MAIN STREET, ONTARIO, CA.

flowers, festooned in leafy plant and growing vine, and enshrined in the hearts of a contented and happy people, whilst around these towns are scores of beautiful homesteads, constantly multiplying and ever increasing in attractiveness, embowered in orange groves and encircled with ornate evergreen hedges, that to be appreciated must be seen. Some may be tempted to ask whether the days of miracles are past. It really seems that such a complete transformation could not possibly be effected in the short time named. But the facts are exactly as stated, and the circumstances are now matters of history. From a howling wilderness to a smiling garden ; from a bare, deserted, barren plain, to populous, thriving towns, with all the accessories of the higher civilization—and all in the short space of five years ! verily, this is Wonderland.

CITRUS CONDITIONS.

For the cultivation of the orange, the lemon and the lime, the best authorities state plainly and pointedly that Ontario not only has no superior but NO EQUAL in America. if indeed in the world. This is on the authority of such able and experienced Horticultural Journals as the *Pacific Fruit Grower*, the *Rural Californian*, etc.; and expert, scientific horticulturists who have devoted their lives to orange culture. Such testimony is valuable and flattering in the extreme, but what is of still greater value—the proof is now forthcoming from the Ontario groves themselves. These young groves, free from all manner of smut and scale; bright, clean and beautiful, have produced this year from $300 to $500 per acre, with trees less than FOUR YEARS OLD! Some three-year-old trees have produced a box each of perfect oranges, about what would be expected in Florida, for instance, from trees eight to ten years of age. From such facts it can easily be gathered that the Ontario land is intrinsically the most valuable in America. In addition to citrus fruits, nearly all deciduous and other fruits do remarkably well—grow luxuriantly and bear early and heavily—such, for instance, as olive, peach, apricot, guava, prune, pear, apple, persimmon, plum, raisin and wine grape, etc , etc

THE WATER SUPPLY

of Ontario has frequently been pronounced its grandest feature, and it really is so. Its purity is absolute, and its quantity is simply inexhaustible. From the engravings presented herewith, some idea may be formed of the surface flow from the Crystal Mountain Streams in San Antonio Cañon, which foam and fret as they sweep along the gorges, roar over cataract and cascade, and plunge down precipices, forming picturesque water-falls and here and there deep pools, shaded by overhanging rock or leaning tree, and thickly flecked with shining Trout, such as fill the hearts of disciples of the gentle Isaak with admiration, and set them all aglow with excitement. Of this vast supply all is now permitted to go to waste. In Southern California " Rivers run bottom side up." So a tunnel was run under the Cañon with a success that was astounding to all save the Engineers, who knew perfectly well what they were doing and what they might expect. Water was struck in 2000 feet, and before 3000 was reached it rushed in with such violence as to carry in great boulders and sweep out the workmen. The tunnel has now been arched and cemented, and over its level bottom rushes a perfect flood, that can be doubled or trebled at will should use or necessity ever arise for further supply.

THE CLIMATE

of Ontario is one of its greatest and most alluring attractions. Dr. Widney and Dr. Lindley of the *Medical Practitioner*, two leading physicians of the State, pronounce the upper end of the Colony, now known as San Antonio Heights, the

HOTEL AND GROUNDS, ONTARIO, CAL.

Sanitarium of the Pacific Slope. Here all bronchial, catarrhal, lung, asthmatic and pulmonary troubles of whatever sort are relieved so far as climate can accomplish that end. Hundreds of people have been benefitted and scores of lives saved by a short residence in this paradise for the afflicted. The elevation of these Heights is from 2000 to 2400 feet, and the view from them beggars description. Once beheld it will remain an inspiration for all time. Before the eye is spread out a perfect panorama of rugged mountain and sloping valley, emerald-clad foot hill and flower-carpeted vale; towns, hamlets and orange groves; the Pacific Ocean in the distance glowing like a mirror under the flood of sunlight, with her turreted islands sleeping peacefully upon her warm bosom. No lover of the beautiful would ever regret a thousand mile journey for ten minutes upon these Heights on a perfectly clear day with no fog upon the ocean. The vision stamps itself upon the memory, and there abides a pleasure to the end of life. And if so much pleasure is to be derived from so short a visit, what must it be to reside there, in a comfortable home, amid such peerless surroundings, with an Eectric Railway to carry you from your very door to two towns and the depots of two transcontinental lines of railway in a few minutes? For San Antonio Heights is laid out on lines of beauty, in three-fourth acre lots, with pressure water, in cast-iron pipes. It is a spot to enrapture the most phlegmatic, and will be the ideal residence town of the great Empire of the West.

THE IMPROVEMENTS

in the Colony are, for its age, remarkable. Two hotels of the value of $50,000 each, besides a half dozen minor ones; one brick block extending from street to

STATE BANK BLOCK, ONTARIO, CALIFORNIA.

street, $55,000; another now approaching completion. $30,000; a perfectly appointed brick livery and sales stable, $10,000. These chiefly in the south town. Numerous other brick blocks of lesser note in both towns. The more expensive only have been enumerated. College, $20,000, with an endowment of $200,000; four public schools; two costing $10,000 each; three handsome churches completed and two others being commenced; three railway depots, for Ontario is crossed by two great Transcontinental Lines—the Southern Pacific and Santa Fé—besides having a local line (the Chino Valley) and an Electric (the Ontario and San Antonio Heights). Of private residences nothing need be said; they are, of course, everywhere. Of expenditures by the Ontario Land Company, $200,000 have been put in water pipe, underground and OUT OF SIGHT. There are no open ditches to invite malaria or breed disease. Where else has a Land Company spent such a vast sum to insure perfect health to an entire community throughout all future ages? There are EIGHTY-TWO MILES of the best stone pipe now laid for irrigation, and FIFTEEN MILES more of cast and wrought iron for pressure in the towns. These are but samples of the improvements, but they are sufficient to indicate what is going on, and they speak in unmistakable terms of the character and enterprise of the people. In a word, Ontario is an example to the world to-day of what can be accomplished by wise foresight and well-directed effort, in a region where Nature has been more than bountiful with her wealth, and simply lavish in all that goes to make life pleasant and enjoyable. And yet her progress has really but just commenced. Her future is beyond the power of pen to depict or imagination to conceive. No brawling saloon can disturb her peace, for these dark blots upon the Country's fair escutcheon are absolutely ruled out of Ontario by a stringent prohibitory clause in every deed. Her triumphs are the triumphs of morality. Her progress is the progress of science, of education and of all the arts of peace. The FIRST STONE laid was that of her college, hence she was founded upon the rock of truth, morality, intellectual culture, and liberty—as portrayed in the teachings of her foremost educational institution. Is it any wonder that she has flourished? and who can doubt that she will continue as she has begun? only at a constantly increasing ratio, as a snow ball gathers in weight and dimensions from every additional revolution. Ontario is a child of destiny. Her future is as assured as the eternal mountains by which she is surrounded and protected; or the great Sea whose tonic breath, divested of every atom of moisture by its inland journey, has done so much to give strength and bloom to her youth. Her activity is but the murmur of the tread of ages yet to come—the faint sound of the march of the foot-fall of a destiny that shall shine as the stars and on the outstretched finger of all time sparkle forever. And here, truly, if anywhere beneath the sun, the citizen is assured of his inalienable constitutional rights of life, liberty, and the pursuit of happiness—and assured of them too under kindly skies, in a healing atmosphere that is the very Balm of Gilead, and amid scenic beauty, tropical growth, cultivated society, and such sense-charming and soul-satisfying surroundings as to leave little to be desired. Nor can this garden of the Hesperides ever become old, for the bloom of perpetual youth is in her life-giving atmosphere, her healing sunshine, her fragrant groves and aromatic plains. On her lofty mountain heights the snows never melt, but in the enchanting valley of this Land of the Afternoon, bathed in a flood of slumberous sunbeams, the Rose never fades.

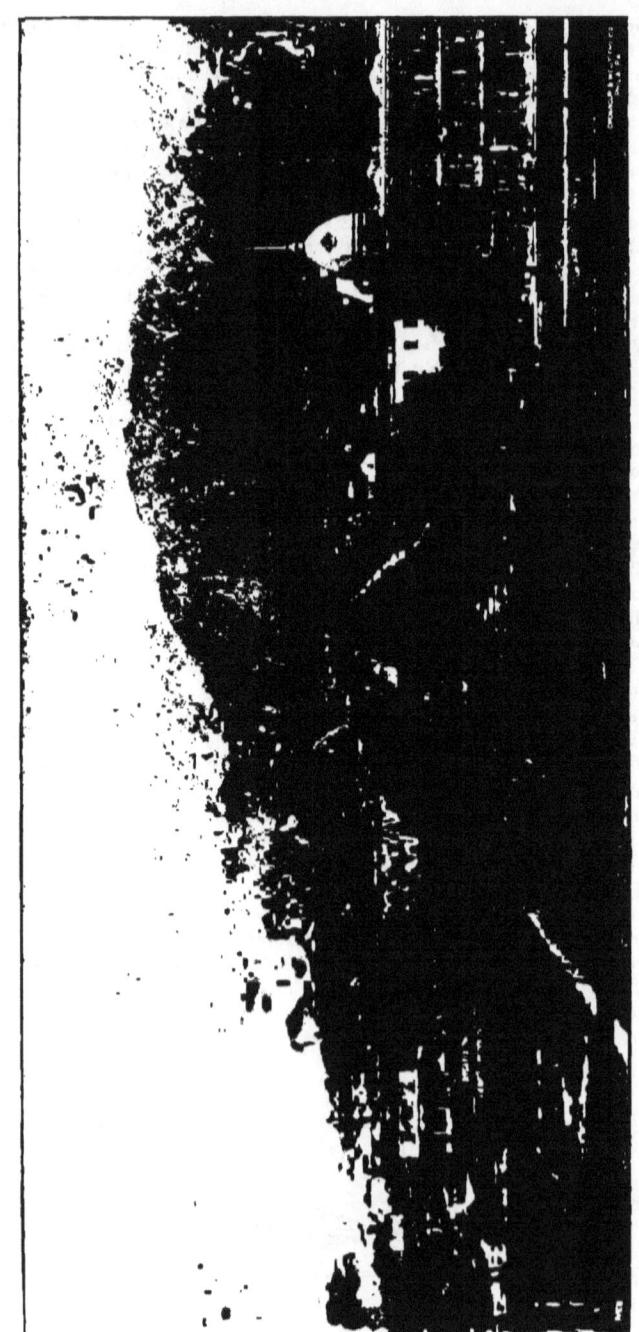

EUCLID AVENUE, ONTARIO, CAL.

COLTON, CAL.

THIS article on Colton, and the villages and country more or less tributary thereto, is necessarily brief. It is inserted in this volume by the Colton Board of Trade, and is reliable, except that it will underrate rather than over-estimate her natural and acquired advantages, as a home, a resort, and as a desirable business centre.

The town is incorporated, has a population of about twenty-five hundred, and an elevation of nine hundred and sixty feet above sea level.

Colton is situated in the centre of San Bernardino valley, which is ninety miles long from east to west, has an average width of about fifteen miles, with numerous small valleys tributary to it. This valley is at once one of the largest, most productive and best watered valleys in Southern California. The two great transcontinental lines of the Southern Pacific System, and the Atchison, Topeka and Santa Fé cross at Colton. We are fifty-eight miles east from Los Angeles, and one hundred and seventeen miles north from San Diego. Our close proximity to the seaports of San Diego, San Pedro, Santa Monica, Port Ballona and Santa Barbara and our direct connections with each of them, renders our maritime advantages excellent. The soil around Colton is largely a disintegrated granite, carrying alluvial deposit, rich in vegetable mold, and some sand. It is very porous also, water percolating freely, making a soil that is dry and spongy, and at the same time capable of retaining moisture almost an incredible length of time. Vegetables of every variety are consequently of remarkably quick growth, and possess the finest qualities of flavor peculiar to their kind.

A very slight difference in elevation determines often the kind of vegetables, as well as the varieties, and fixes by immutable laws the kinds of fruits and cereals that shall be raised. To illustrate: potatoes grow luxuriantly anywhere, but in the market you are asked whether you prefer the mountain or the valley grown potato. It is also found best to grow different varieties of grapes in different altitudes. Bananas and dates seldom come to maturity here but may be found growing in many localities.

To attempt to name the forest, fruit, vegetable, cereal, floral and shrubbery growth of this valley would be quite impracticable. The catalogue of fruits and cereals, semitropic and temperate, is complete, and the names of our flowers and shrubbery are legion. The citrus fruits of Colton Terrace rank equal with the very best grown on the American continent.

The fruit being absolutely free from fungus and scab, the skin smooth, varieties complete, size above the average, the flavor of each of at least six varieties decided and perfect, together with the cleanliness and health of the trees, their luxuriant growth and abundant and unfailing yield, renders the culture of this crop both pleasing and interesting and among the most lucrative pursuits of the valley.

Our wine cellars represent every variety of sweet and sour wines; our raisins command the best market prices, and our table grapes can not be surpassed on the slopes of Sunny Italy, nor on the vine-clad hills of Spain and France.

Peaches, nectarines, apricots, prunes, and loquots are paying crops, the yield being sure, and the fruit the most luscious. Choicest cherries and apples are grown in the foothills.

Home-seekers and invalids have but one question usually to ask, and that is in regard to climate and water.

Our climate is dry, equable and mild. There is little evaporation from the soil and very little decomposition of vegetable matter. There is no malaria whatever, and no well-defined cases of cholera infantum. The rains fall as the gentlest April showers of the North, there being scarcely any wind whatever, and only occasionally a very little thunder or lightning.

Sunstrokes are unknown in the valley on account of the extreme humidity of the atmosphere. Our nights are always cool in summer. Our winters are very delightful northern springs.

Nothing can be more pleasing to the traveler en route to this country than to rise in the morning of some cold winter day and in an hour to glide from the regions of eternal snows into this valley of perpetual song, sunshine and flowers. It is not in the memory of the oldest inhabitants that it has snowed in Colton. An abundance of pure water is obtained in Colton from wells at a depth of from forty to seventy feet, and also from springs and artesian wells from two to four miles distant, from which it is conducted to us principally in pipes that have a descent of from forty to sixty feet to the mile; thus insuring sufficient pressure, and preserving the water pure and clear as if dipped from its sparkling sources. At present water is supplied by corporate enterprise, but the city shall undoubtedly have adopted a system of water supply for its citizens by the time this is in the hands of the reader that will surpass any system in the state, one that can supply artesian water to its consumers in great abundance and at nominal cost.

It will be seen by the following industrial enterprises that Colton is fast becoming a manufacturing, as well as a railroad centre. In fact there are more laborers now and constantly employed in Colton than find work in any other town of its size in Southern California.

Slover, or Marble Mountain, is a solitary peak which adjoins the southwestern corner of our corporate limits, rising to the height of six hundred feet and is a solid mountain of marble and onyx. The marble is from pure white to jet black and of excellent quality, as is attested by its use in the construction of the finest and largest structures in the state. Both the onyx and marble admit of the most perfect polish. The marble works, at the southern base of the hill, employ about one hundred hands and are now doubling the capacity of their machinery.

The lime kilns on the west side turn out eight thousand barrels per month of the finest lime in the state, and are increasing their capacity. The supply of pressed brick is not equal to the demand upon our kilns. The Southern Pacific Company alone having given the proprietors an order for five millions of brick.

Our planing mill and sash and blind factory is complete and employs from forty to sixty hands, turning out as fine doors, blinds, sash, molding and bric-a-brac as are used in our region.

A pipe factory is just recently located in our midst. The buildings are in process of erection, and the enterprise promises much for the city in the line of manufacturing. They will employ about forty hands.

Our cannery is a pride of the city, being one of the best regulated and most complete in the country. It has a capacity of from fifteen thousand to twenty thousand cans per day, and employs during the busy season from two hundred and fifty to four hundred hands daily. The goods of the Colton Packing Company are found on the shelves of the best grocers in the land. The company can not nearly supply the demand made upon them.

Our barley crusher, which is operated by water from an irrigating canal, has a capacity of twenty-five thousand pounds per day. The occupation of fruit culture is very lucrative, and is the basis of seemingly high prices realized upon real estate in some portions of our valley.

It passes without comment that our deciduous fruits and berries are among the most luscious grown in the United States.

As a single illustration of the productiveness of blackberries, permit us to state that from nine rows of bushes, that were planted between rows of peach trees, each forty rods long, over four and one-half *tons* of berries *were sold*, many that were picked were not accounted for, and many dried on the bushes.

The business facilities of Colton are excellent. We have fifty-eight daily trains running toward six different points of the compass; eighteen of them are mail trains. We have an electric car line in process of construction to San Bernardino. The California Southern Motor Railroad Company, with headquarters in Colton, have a line to San Bernardino, a d contemplate in their system an extension to Arrow Head Hot Springs, Highlands Hot Springs, Mentone, and the eastern end of the valley; also a line to the Northwest and West connecting Rialto, Ettiwanda and other towns with Colton; and toward the South, which is building, a line which will give Riverside and a number of other towns a second direct communication with Colton.

A narrow gauge is now building into the foothills for the purpose of transportation of wood, lumber, bark and ice. We now have over fifty cities, towns and villages within a radius of sixty miles, all connected with Colton by rail. These new railroad facilities already begun will place many more towns in communication with us. Colton is one of the very few towns in Southern California in and around which the Southern Pacific owns a large interest and where that company is doing much to advance and prosper the town.

Colton is a shipping point and is the fifth largest in freight receipts on the line of the S. P. R. R. in this state. Large stock and feed yards have recently been erected here by this company.

This means a great deal to the grain and hay producers of our vicinity. Cattle are brought here from Texas, New Mexico, Arizona and Sonora, fed here and reloaded for San Diego, Los Angeles, San Francisco and the North. Machine shops have been talked of by the S. P., but nothing has been done as yet toward locating them here; the number of lines now radiating from this and adjoining towns will soon make it necessary to have a source of supply and repair in this immediate region. Many of our hotels, restaurants and boarding houses are good and their capacities are usually taxed to the utmost. A hundred and fifty room hotel was begun during June 1888, and as the projector is a practical hotel man of large experience and means, we look forward to the time when we shall be able to say our accommodations are as complete as are found in this part of California. Electric light wires are now spread over our city and these beautiful lights are fast being considered a necessity with us. We have an extensive wholesale grocery house, a solid national bank, a daily and weekly newspaper, some good general stores, nearly every line of commercial trade represented, and two building associations. The Union Ice Company of Southern California have their headquarters here.

Our educational, religious and social advantages are superior, as our fine graded school, with a building costing fifteen thousand dollars, numerous churches, W. C. T. U., Y. W. C. T. U., I. O. G. T., I. O. O. F., free reading room and public library attest.

And now a word as to enterprises that would prosper here. This is a headquarters for fruits. An evaporator would do well here. The business of curing stone fruits, raisins, figs, and berries would prove very lucrative.

Hides and pelts can be brought here from the vast grazing lands of Texas, Sonora, New Mexico and Arizona, and from the cities of San Diego, Los Angeles, and every other city and town in Southern California almost directly. Thousands of cords of oak and hemlock bark are wasting on the mountains not over fifteen miles from here. This then would make a good centre for a tannery. We have already hinted at the large number of cattle and sheep that will be unloaded and fed here. Why, in view of our railroad facilities, would not this be a capital place for a slaughter house and cold storage institution? An ice factory is needed. Pottery and tile works can find good clay here and excellent market. A dairy and poultry ranch would be one of the most lucrative businesses here. A good carriage and wagon factory will find the best of material for woodwork at their very doors. Hundreds of thousands of fruit boxes are imported here year after year; the timber is near by, and the demand for such boxes is every year increasing. Any information in regard to Colton or surroundings will be furnished gladly by addressing the president or secretary of the Colton Board of Trade.

PALM VALLEY.

At the eastern base of the Great San Jacinto mountain, in San Diego county, California, and just over the San Bernardino county line, lies the noted Palm Valley, noted for its unequalled climate; noted for its magnificent scenery; noted for its fertile soil; noted for its pure snow water; noted as the

EARLIEST FRUIT REGION IN CALIFORNIA.

The valley is about 50 miles from Riverside and 120 from Los Angeles, and communication with the cities and markets of the world is had by means of the Palmdale railroad, which connects the valley with the Southern Pacific at Seven Palms.

The valley is protected on three sides by high mountains, and the scenery is grand in the extreme. The great San Jacinto on the west, towering up out of the valley over 10,000 feet in the most abrupt ascent in the world, effectually shuts out all possibility of heavy north winds, and such a thing as *fog has never been seen in the valley*. Frost, too, is unknown, and the most tender plants and all tropical fruits mature here in perfection.

The winter climate averages, both day and night, from 7° to 10° warmer than at Riverside or Los Angeles, and the result of this condition of temperature is that every variety of fruit matures here from one month to *six weeks earlier than at any other early fruit district in the entire state*.

This year (1888) watermelons were shipped from Palm Valley during the month of May, and those first shipped from any other point came from Lodi, and arrived in San Francisco July 2d. Ripe Mission grapes were received in San Francisco from Palm Valley on June 18th, and the earliest Missions are not expected from any other place before September 1st. Many other facts of early productions could be stated would space allow. The rapidity of vegetable growth is astounding. Beans have been known to grow six inches in four days from planting; grape cuttings show a growth of ten feet in four months; a cypress vine has grown six inches in a day, and fig trees have grown three feet in a month. The result of this most rapid and early maturity is that producers can realize immense prices for their crops, being able to market them at times when there is absolutely no competition.

The Palm Valley Land Company, composed of well known San Francisco and Riverside capitalists and horticulturists, after proving to their own satisfaction all these facts, purchased, during the latter part of 1887, all the best available lands in the valley, and have spent large sums of money in the development and improvement of the property. They purchased valuable water rights and have completed a water system as extensive and perfect as that of any plant in the state. Over 12 miles of stone-walled ditch have been constructed, besides the various flumes and open earth ditches which complete the system.

The water is pure and plentiful, and each purchaser of land becomes a shareholder in the Water Company, thereby acquiring a perpetual water right.

The company has built the railroad connecting the valley with the S. P. R. R., and have planted *160 acres to Navel oranges*. This great orchard, THE LARGEST NAVEL GROVE IN THE WORLD, adjoins the town site—Palmdale—and is held by the company as a permanent investment. This valley is the natural home of the orange, and the orchard is in a most thriving condition.

Palmdale, the terminus of the railroad, is a beautiful location, from which a fine view is had of the entire valley. The streets have been graded and lots will be offered for sale about November 1, 1888.

PALM CANYON.

At the upper end of Palm Valley the mountains close in, until apparently there is no further advance to be made in that direction. A short walk along a well-known trail over the hills leads one to a cañon, where a scene bursts all at once upon the vision, the like of which cannot be found elsewhere in all the United States. The rains of ages have washed the soil from the surrounding mountains, until it has accumulated in the bottom of a narrow valley, forming an admirable bed for the growth of all sorts of vegetation. The growth here seen is a most effective witness to the fertility of the soil and the mildness of the climate of Palm Valley. As a sanitarium and resort, Palm Valley will soon be famous.

Colony tracts, in size from 5 to 20 acres, are now on sale, and full particulars, maps and catalogues can be had of the general agents,

BRIGGS, FERGUSSON & CO.,
314 California St.   SAN FRANCISCO, CAL.

SOUTH RIVERSIDE.

THE QUEEN COLONY OF SOUTHERN CALIFORNIA.

SOUTH RIVERSIDE Town and Colony are situated in the southwest part of San Bernardino Co., California, in the famous Riverside Orange belt, on the California Southern Railroad (the Santa Fé system), fifteen miles southwest of Riverside and forty-four miles from Los Angeles. The Pomona, South Riverside and Elsinore Railroad is already graded from Pomona to and through South Riverside and grading has also been commenced at Elsinore, and the road will soon be open for business. An extension of this road from Elsinore to San Diego is being surveyed and the probabilities are that the entire road to San Diego will be completed within a year. The Southern California motor road from San Bernardino now completed to Riverside, will be completed to South Riverside within the year. The San Bernardino and South Riverside Railroad has recently been incorporated and in due time will be constructed. Still another railroad to South Riverside is proposed. Everything points to South Riverside soon becoming one of the most important railroad centres in Southern California. The vast resources of minerals and products of the soil contiguous to South Riverside will make a large business for railroads. The Town and Colony, but a little more than two years old, have had a phenomenal growth. They comprise a growing community and there is room for all seeking to establish themselves in homes or business. A new place with such a record and such prospects has much to concern the general reader and claim the attention of the home seekers and careful investors.

All things considered, this tract is unequalled in the great orange belt, it is near enough to the ocean to get the refreshing breezes, far enough inland, separated by mountains, to escape the fogs and mists of the coast, while well protected from northers in winter and scorchers in summer. The greater portion of this tract is a gentle sloping mesa, choice foot-hill land, from ten to fourteen hundred feet above sea level.

The air is pure and invigorating, sunshine and clear weather most of the year, cloudy days the rare exception. Some who have long suffered from lung and other troubles have received great benefit; cases almost hopeless have been entirely cured; the general effects of the climate are easily seen but not so easily described. South Riverside will become a great health resort. In the Temescal valley, part of South Riverside Colony, are the White Sulphur Springs, warm and cold, well known for their curative properties.

Southern California has gained a world-wide reputation for its winter climate, and is rapidly becoming a popular resort for all who would escape the rigors of winter weather east of the Rockies; the time is not far distant when it will be quite as favorably known as a summer resort, especially on the beach, and mountain streams. Strange as this may seem at first it can be verified. Summer in Southern California does not mean hot days and nights as many think. As a summer and winter resort, California has no equal; it is the world's sanitarium, pleasure ground, El Dorado.

This great fruit land tract comprising 15,000 acres, including the Auburndale subdivision hereinafter mentioned, invites capital workers, capital in brains, active capital; here is an open field where well directed effort will have its reward.

These fertile lands have long waited the magic touch of water and the husbandman's skill and thought. Now the change has commenced, in a short time thousands of acres in trees and vines will yield their golden fruitage. The land is divided into tracts of two to 10 acres each.

The olive of the Orient will flourish here, and the time is not distant when the olive oil industry will become one of the most important ones on the South Pacific coast. Orange culture in this valley has already proved a great success; and it is predicted that South Riverside will add to the well known reputation of this orange belt, and fairly deserve the present appellation, "Queen Colony of the Golden State," being the natural home of the citrus family, having the rarest climatic conditions. It must become an active centre for fruit growing and other business, a thriving place of wealth and influence.

These lands are not low-priced, but they are cheap in view of their productive worth and the rapid advance that must continue as they improve, and come more and more in demand. The soil is rich and deep, suited in every way for horticulture and capable of untold wealth of products. These orange lands are worth from $150.00 to $300.00 per acre, including water.

Water is the life of California land. Ample water for this colony is brought in large pipes from mountain streams, scienegas, springs, artesian wells, etc., piped to the land ready for irrigation and domestic use. Water rights are sold with the land, giving owners the right and use of water without further cost.

There are several thousand acres of this land especially adapted to the growth of oranges, grapes and other choice fruits. The reputation of this section for orange culture is well established and widely known. The Washington Navel is a great favorite; has passed the test of quality and commanded the highest prices in some of the best markets of the world.

The Town and Colony of South Riverside are unlike any other in the State for novel design and beauty of location. The town site is in the form of a circle and all leading streets of the whole tract centre there; blocks have twelve lots each 50x150 feet; the streets are broad and well graded, and will be lined with a variety of evergreen trees; four beautiful parks are designed within the city limits; the grand boulevard is one hundred feet wide and three miles long, thus making a beautiful drive around the circle city of the citrus belt. Young as the town is, it has schools, churches, hotels, bank and business blocks, handsome dwellings, and a fine union railway station, buildings all in all that would be a credit to any place. These improvements, amounting to many thousands of dollars, are modern and substantial.

The famous Magnolia Avenue of Riverside is now extended across the South Riverside tract. When completed this will make a continuous avenue one hundred feet wide and fifteen miles long. It will be one of the finest driveways on the continent.

South Riverside will become a live manufacturing point for fine crockery ware, piping, tiling, building material, etc. Water lime of excellent quality, carrying a large per cent. of cement, is found near by; granite, porphyry, and gypsum are found in abundance; coal of good quality, two good veins having been discovered, promising rich results, solving the fuel question, and determining one of the best locations in the State for manufacturing enterprises. Gold, silver, copper, other ores, and the richest tin mines in the world are but a short distance from the town. The title to the tin mines has just been determined by the United States Supreme Court and work on the mines has been renewed. The great variety and quality of natural resources and the many advantages here await only time and money to turn them into a thousand and one uses. The day will come when the place will be well known for its rich mineral and manufactured products. Arrangements for a sewer pipe factory are now being made.

It is rare to find such a combination of natural advantages; capital can find here opportunity for paying investment; labor will seek this locality for steady employment at good wages; the outlook to the new-comer will improve the more he investigates. At first it may seem overdrawn but it will require a short time only to see something of what South Riverside will be the next ten years.

These lands are sold at prices much lower than elsewhere for the same quality of soil and similar climatic conditions. The terms are extremely favorable, only about one-fourth cash and the balance drawing seven per cent. interest, payable on or before ten years at the option of the purchaser. The property will continue to advance in value as improvements go on, and the demand for choice acreage increases. The place is not an untried experiment. Worth and merit are at the bottom of it all.

Some of the many claims that will concern home-seekers and careful investors are fertility of the soil, healthfulness, being near enough the sea for comfort, far enough inland for finest fruits, the orange and olive, lemon and lime, grapes, apricots, fruits of all kinds in abundance. Size and quality will not be excelled. Grand scenery, fine climate, beautiful location for homes, railroad facilities, within easy reach of large towns, a community of live workers bound to succeed.

Such, in brief, is the new place to which you are invited. Come and see for yourself, look the ground over, if it suits, invest, take hold and help along public and private enterprises, become an active citizen, share in the present and future prosperity of the "Queen Colony," "Gem of the Orange Belt," South Riverside, Southern California.

For information address

SOUTH RIVERSIDE LAND AND WATER COMPANY,

SOUTH RIVERSIDE, CALIFORNIA.

ALESSANDRO.

The Tenderloin of Southern California.

Wonderful Natural Advantages and Possibilities Offered at Nominal Outlay.

ALESSANDRO, Cal., January 12.—It is but natural that people with moderate means should look about for a location in this most favored region where they will enjoy the reward of development and growth.

There are to-day thousands of people ready to come to Southern California, and the first impulse of these is to make their future home in some of the larger cities, never considering the possibility of securing anything desirable in either town or acreage property at a moderate figure. All the cities in Southern California have had "booms," which the new comer will find a very expensive luxury. And it does not pay to travel through this country with a family any length of time hunting a location. The writer, who has enjoyed special advantages in investigating the true merits of this State, does not hesitate to say that a large portion of the same is far from being the Garden of Eden generally represented in cheap boom literature, and those who are foolish or unfortunate enough to tie to them will ere long return East, poorer, though wiser, men. On the other hand, Southern California contains a few strips of land, charming valleys mostly, which, under cultivation, will produce abundantly everything raised under the sun, and offer every settler an independence and fortune within the shortest space of time. And it is to this region that my attention has been mainly directed. It appears at these points as if natural advantages had vied with each other to excel and to shower eternal wealth and happiness upon those who would be the lucky ones to aid in their development.

While many thriving towns of this Bonanza Belt are already favorably known to the readers, it was not until a few days ago that an accident permitted me to discover the tenderloin of Southern California, just thrown open to settlement, the centre of which forms the newly laid out town of

ALESSANDRO.

In this age of paper towns and real estate craze, it is not always easy to discriminate between the embryo California towns with or without a future, and only the experienced eye can successfully penetrate the mystery and brush away the clouds of uncertainty hovering about these enterprises. And while many new towns of Southern California will never be heard of beyond the lithograph establishments, a few will become cities within an exceedingly short space of time. And prominent among these will be charming Alessandro.

This new town is the creation of the sterling firm of Messrs. French, Packard & Rockwell, of Pomona, Cal., whose far-sightedness and sound judgment led them to acquire by far the choicest and most superbly located tract of land, comprising 10,640 acres in the richest and most fertile portion of Southern California—the San Jacinto plains—the products of which will find a ready market in this future metropolis.

Alessandro is located on the main line of the California Southern Railroad, seven miles southeast of Riverside and 100 miles north of San Diego. The scenery surrounding it stands without a peer. To the east, the majestic snow-capped peaks of San Bernardino and San Jacinto, with their range of mountains, form an imposing barrier to the cold winds and lend a grand background to the beautiful scene unfolded here, and one cannot resist the feeling that Madam Nature had carved out her favorite town-site right at this point and deposited at its very doors all the wealth of her luxurious larder. The land surrounding and tributary to it is of the most fertile and adapted to the profitable cultivation of all kinds of fruit, cereals, etc.—in fact, everything grown everywhere.

The town of Alessandro comprises only 240 acres, the liberal-minded and conscientious owners preferring to eliminate as far as possible every vestige of unwarranted speculation and to permit the new town to grow up upon its merits only. A large amount of money has already been expended in improvements and it is the intention of the gentlemen interested to show their unlimited faith in this new town by putting in their own money first. Already the coming prosperity is evinced on all sides, and the busy sound of the hammer around here seems to have a particularly substantial ring. A handsome hotel has been completed and a fine depot building erected by the California Southern Railroad. Several other buildings are up, as well as a store building. Four fine brick blocks have been contracted for and will bid a cheerful welcome to the new settler within a short time. Two beautiful parks are being laid out, which will be covered with prolific semi-tropical vegetation. A grand avenue—Majella avenue, five miles in length and 100 feet wide—is being graded, and when adorned with a double row of graceful pepper trees will offer one of the most exquisite drives in America. Thanks to the enterprise of the owners, water for domestic purposes has been already provided by an expensive system of pipes, and every home will find at its very doors the purest, clearest water in abundance, a matter which cuts a very important factor in Southern California.

To encourage superior school buildings, the owners have agreed to set aside one per cent. of the total sales of this extensive tract, which will amount to quite a large sum, when the immense area is considered. Two and a half blocks have also been donated for church purposes, and I learn that all necessary land and even some money will be cheerfully given by the public-spirited owners for like purposes. Indeed, there is not a town-site which has at its back the broad-gauge spirit and liberal bank account of the gentlemen who hold the reins of Alessandro, and the people who give this promising enterprise their preference will never find reason to regret their judgment.

To the tiller of the soil, the Alessandro tract appeals pre-eminently. Those who understand California know that only a combination of four indispensable factors insure the enormously profitable crops so often read about. These are climate, soil, water and elevation. And all of these are found at Alessandro, as if made to order.

The climate is simply perfect. Entirely devoid of frost, and sheltered by the mountain ranges from disagreeable winds on one side and the unwelcome fogs on the other, the atmosphere is invigorating and bracing. and, perhaps, represents better than any other section I visited, the ideal climate of California.

The soil is of the very best. The railroad cuts diagonally through the tract, the greater portion of which lies to the east and contains what is called moist lands, upon which water can be had in abundance almost anywhere at a depth of from five to fifteen feet of the surface. Here all kinds of agricultural products can be raised in abundance without irrigation. The dry lands are on the west side of the tract, and will require irrigation. The owners of the town site, who have investigated the matter very closely, have abundant evidence of the existence of artesian water upon this tract, and are now boring an artesian which will enhance the value of the present low prices very materially. On the other hand, abundant water can be had upon every foot of ground by digging wells sufficient to irrigate small tracts, which, owing the fabulous returns, are the rule rather than the exception in California.

The elevation is particularly fortunate. Statistics show that the best results in the raising of oranges and fruits, generally require an elevation of between 1,000 and 2,000 feet, which precludes the possibility of frost. Alessandro is situated 1,500 feet above sea-level and in this as other advantages, readily distances many less fortunate regions.

With the natural advantages equal if not superior to the most noted sections of Southern California, coupled with the enterprise and unlimited capital of a wide-awake corporation, who would doubt the future of Alessandro? While the property just placed on the market is only held at $25 to $125 for town lots 50 x 150, and $25 to $130 per acre for acreage property, it is a safe prediction to look for a large advance within a few months, just as soon as the contemplated extensive improvements can be gotten under way.

Want of space precludes more extended mention to-day, and I take pleasure to refer those interested to the Alessandro Land and Water Company, whose efficient officers, Hon. John L. Means, of Grand Island, Nebraska, president; Charles French, vice-president; G. E. Ross, secretary, or People's Bank, of Pomona, treasurer, have handsome offices at Pomona, and will cheerfully furnish all information desired regarding this favored spot.

I can only add that I consider this tract of land and the new town-site of Alessandro far ahead of anything I have found in Southern California as offering a veritable bonanza to the man of only moderate capital.   ERTEL.

THE LAKE COLONY,

SAN DIEGO COUNTY, CALIFORNIA.

A place that should not be missed by any tourist is the Elsinore Lake—the only lake in Southern California. It is a lovely and placid sheet of water, two by five miles in extent, located half way between the two important cities of Los Angeles and San Diego, twenty-two miles inland, at an elevation of 1,280 feet above the sea. It is easy of access, being on the main lines of the A., T. & S. F. and S. P. railroads from the east to San Diego; the former road having been operated along the lake shore six years, the latter now being constructed.

The rapid filling up of Southern California during the last four years has brought several thousand people to this unique and beautiful spot, where they have built up homes and towns which are the admiration of all visitors. Man is doing by enterprise and thrift the little that was left undone by nature to make this the most desirable site for a home.

Mr. J. H. Roe, of the *Riverside Valley Echo*, says, in a recent editorial that, "Elsinore with its lake and surrounding settlements is certainly the most romantically beautiful spot in Southern California. We say this advisedly after being familiar with Pasadena, San Gabriel Valley, Arrowhead, Redlands, Riverside and all the seaside resorts on the Southern Coast." This is nothing more than honest appreciation candidly expressed, and the almost universal verdict of visitors, who make the complete circuit of the lake and its surrounding towns. The oldest of these is called Elsinore; the best railroad and agricultural town is Wildomar; the most lovely and commanding site is Lakeland; while the productive mineral belt is centered at the Chaney Coal Mine. Space forbids the detailed description of these places that their importance would warrant. We can only hint at the advantages of each and trust to readers of the Handbook to go and see for themselves. They will be well repaid.

WILDOMAR

Is on the line of both railroads; is the business centre of about six thousand acres of rich fruit, farming and grazing lands; is in its third year; shows remarkable and substantial growth, and has an assured future. A cut of its cozy hotel is presented herewith as a sample of its improvements. It contains twenty-six rooms, well finished and furnished, and kept as a first-class country hotel, the comfort of the guests being carefully looked after. This is the most convenient stopping place for visitors to the colony. From here a hired livery rig or a free real estate agent's carriage will convey you along the broad and level Grand Avenue five miles to

LAKELAND.

Here the wise tourist will turn up one of the broad avenues of this charming though incipient villa tract, look at its liberal lots commanding a view of the entire lake and surrounding valley, and if he does not secure a lot, he has stronger resisting force than the writer can boast. He can not help thinking that he wants one of them, like Mrs. C. B. Jones, whose artistic taste selected the site and helped to found Lakeland, "to think from," if he never gets there to live. Much as we would like to linger, we must continue along Grand Avenue and, leaving it and the lake, visit the now famous

CHANEY COAL MINES,

Located four years ago, by the gentleman whose name they bear. Having been slowly developed until the worst of the "coal famine" of the last winter, they suddenly commenced to furnish two carloads per day. This was a limited supply but it brought the coal "corner" to time and saved the people much money. The field is now well explored and the supply demonstrated to be almost inexhaustible. The quality is good, improving as the entry progresses. The vein varies from five to eight feet in thickness. The S. P. R. R., now building, will pass right by the mine and furnish cheap transportation. The Santa Fe have mooted the idea of a branch from the mines past Lakeland to Wildomar. The sewer and water pipe works, now in operation close to the mine, use about ten tons of coal per day and will need more in future. Other manufactures are contemplated which will greatly increase the local demand. There is an opening here for some live men who can buy a fourth interest at a low price and assist in the handling of the coal, or construct and operate works to utilize the extra fine quality of fire clay that is found in the mine above the coal; or in working some of the many other minerals in the vicinity, with the aid of cheap fuel.

Continuing our pleasant and instructive drive we come around to Elsinore, prosperous and pushing. Again skirting the lake several miles we pass Elsinore's railroad station, and thence through the grain and fruit farms, home again to Hotel Wildomar. For further particulars the reader can obtain bird's-eye views, maps, etc., free, from D. M. GRAHAM and WM. COLLIER, of Wildomar, Cal. They are the founders of the towns above mentioned and part owners of the coal mine; and will be pleased to answer letters and personal applications.

MONTEREY, CAL.

A GREAT WINTER RESORT.

The Celebrated HOTEL DEL MONTE and its Seven Thousand Acres of Pleasure Ground on the Pacific Shores—One of the Most Magnificent Seaside Establishments in the World—MONTEREY and its Surroundings—A Royal Resort in a Romantic Region—Interesting Items.

With her natural resources known to her own people it is a singular fact that until less than a dozen years ago very few people, except those who had visited California themselves, thought of her as a resort or winter home, a region in which to regain health, or in which to find pleasure, rest and recreation. In the spring of 1880 an event transpired that marks an epoch in the annals of the state's history, for from that time on thousands of people have heard in all parts of the world that California has in her possession "The Queen of American Watering Places." The event which made this known was the opening of the Hotel del Monte and resort at Monterey. Following closely the completion of the immense Pacific railway system which bind the two coasts of America, the opening, not merely a hotel, but of 7,000 acres of pleasure grounds, greater, more costly, more magnificent than any winter resort in the world, the dedication to the public of the Hotel del Monte and its grounds, gave to the Pacific Coast a new meaning in the minds of thousands of people throughout the entire land. Previous to this event little was thought of any part of California or its coast as a winter resort or summer watering-place ; the opening of Monterey marked a revolution in this respect; noted people came from Europe and America, tasted of her pleasures, were enchanted by her attractions, and spoke of them to the whole world.

The accompanying picture affords the reader an opportunity of gaining a little conception of the external appearance of the hotel, with just a little glimpse of the grounds surrounding it. The hotel contains very nearly five hundred rooms and can easily accommodate seven hundred and fifty persons. In furnishing and in interior finish of the hotel throughout, expense seems hardly to have been considered at all, the idea prevailing to have the most artistic and at the same time the most appropriate and durable, giving the effect of real merit and worth. The carpets are Axminsters, Moquets and Brussels; the woods used are San Domingo mahogany, English quartered oak and selected cherry. All the rooms of the house are furnished equally well ; though variety has been sought in different colors, designs and finish.

To those who have never visited Monterey, a description of the grounds and surroundings will be of interest and importance. In other instances we frequently hear of a hotel, standing by itself perhaps on a barren beach or bluff, surrounded by a sandy waste, spoken of as a *resort*. How vast the difference between such a resort and the fair Hotel del Monte, located in its enchanting garden of nearly two hundred acres, with seven thousand acres of forest and sea-coast adjoining ! The traveler visiting the Hotel del Monte alights at the little station house; through the foliage of the large live oaks, pine and cedar, in the distance, he catches glimpses of the beautiful hotel. Proceeding toward the house by carriage or on foot, the park grows more and more picturesque, more enchanting, more surprisingly beautiful. The hotel, conspicuous though it be, is lost from view, it can not occupy but a secondary place in the picture. Under the great, rugged, gnarled oaks have been laid in graceful curves the smooth graveled walks and drives. Approaching nearer to the hotel we see the work of the artist in flower-bordered walks, intricate figures wrought in velvety beds of various tinted flowers, and in the selection and arrangement of various plants and shrubs from other lands and climes, all growing in profusion. Various species of cacti, century plants, prickly pear, and other plants that thrive in the perpetual summer of this paradise and esteemed curiosities in cold countries, add to the interest and beauty of the scene. Beneath the large oaks, hung with long, drooping moss; and around the base of the great pines, laden with cones so large that they seem real curiosities unlike their kind elsewhere, the grass is green and soft, filling the spaces between the beds of rich colored flowers and the smooth walks. In one portion of the grounds is the "maze," a labyrinth formed of cypress hedges, pervaded by foot-paths. To enter is to be lost, and humiliate one's self by calling for a guide in order to escape the intricacies of this curiously wrought puzzle. At a distance from the hotel is an artificial lake, supplied from the Del Monte water-works system and equipped with boats. A feature of the park, some distance in front of the house, are two fine croquet grounds, a lawn tennis ground, and a bowling alley.

When the Pacific Improvement Company had been formed to establish the finest summer and winter resort then known, it was highly essential that before expending the vast sums of money necessary to carry out the project, the most desirable locality be found so far as regards the temperature, rainfall, and other climatic conditions that affect the comfort and healthfulness of the human being. In this respect Monterey has the right of claiming to stand pre-eminent. Statistics prove this. Monterey has for many years been known for its equable temperature. The first capital of California, founded nearly a hundred and twenty years ago by Franciscan missionaries, it has been the cynosure of the coast towns for health, beauty and natural attractiveness, even many years before man had done so much to perfect this garden of the Pacific. The following table, carefully prepared by well-known authorities, whose names might be given, most of the figures being official reports, gives the temperature of Monterey and many other resorts and places:

PLACE.	JAN.	JULY.	DIFF.	PLACE.	JAN.	JULY.	DIFF.
	DEGS.	DEGS.	DEGS.		DEGS.	DEGS.	DEGS.
MONTEREY, Cal.	**50**	**65**	**15**	New York	31	77	46
San Francisco, "	45	66	21	New Orleans	55	82	27
Los Angeles, "	55	75	20	Naples	46	76	30
Santa Barbara, "	56	74	18	Honolulu	59	74	15
San Diego, "	53	70	17	Funchal	60	80	20
Santa Monica, "	54	70	16	Mentone	40	73	33
Sacramento, "	45	73	28	Genoa	46	77	31
Stockton, "	49	72	23	City of Mexico	52	67	15
Vallejo, "	48	70	22	Jacksonville, Fla.	58	80	22
Fort Yuma	56	92	36	St. Augustine, "	59	77	18
Cincinnati	30	74	44				

Many people who have never visited California erroneously imagine that during the "wet season"—so called in contradistinction to the dry months—rain never ceases to descend. This popular error is corrected by glancing at weather tables, which invariably show that during the wet season in California there is not only less rain, but more fair and beautiful days than in any other portion of the United States the same time. Statistics show that the average yearly rainfall at San Diego is ten inches; Santa Barbara, 15 inches; St. Augustine, (Fla.), 55 inches; St. Paul, 30 inches; Mentone, 23 inches; Los Angeles, 18 inches; Monterey, 14 inches.

After a description of the Hotel del Monte and its grounds as a resort, when drawing a comparison between it and many other resorts, which consist principally of a hotel building alone, the most surprising feature left to enumerate and one very acceptable to thousands of guests is the reasonableness of the charges. Hundreds of tourists here in the East testify to this. The rates at the Hotel del Monte are just the medium rates of commercial hotels in cities, and actually about half that charged for the same accommodations at similar hotels elsewhere. This applies to the hotel, and does not refer to the latitude allowed the guest at this resort, where he takes a boat ride on the lake, plays croquet, lawn tennis or billiards without money and without price. From what has been said it is evident that the hotel itself, when classed with other resort establishments, is justified in demanding the highest rates, since every comfort, convenience, and attention is afforded the guests, to be had at the highest-priced hotels of metropolitan cities or other noted seaside resorts. To verify these statements the Hotel del Monte only need refer to tourists in different parts of the land who will testify to these facts. The liberality of the proprietors in this respect even offsets any additional expense that the journey from the distant East may incur, when compared with resorts nearer home.

To the tourist who leaves the ice-bound Atlantic Coast; the frozen streets of Chicago, St. Paul, or other inland cities, in mid-winter, arriving at Monterey after a ride of less than a week, the delight experienced in such a change must be felt; it can not be expressed in words. He feels that at the Del Monte he has found something more than summer weather, summer air, summer sunshine. While Monterey is a delightful summer watering place, thronged by thousands from San Francisco and elsewhere, the summer habitue of the resort does not form an idea of what this "Queen of Watering Places" is to the Eastern visitor, who on arrival can not realize at first that the change is real, the beautiful climate and surroundings permanent. It seems more like a dream.

THE WATER TOWER, CHICO VECINO.

For full particulars, Maps, Pamphlets, etc., address

CAMPER & COSTAR,

CHICO, CALIFORNIA.

FROM NIMBUS KNOB TO THE SACRAMENTO, CHICO VECINO.

SIR JOSEPH HOOKER OAK (29 feet in circumference), CHICO VECINO.

What is Necessary for a Model Home?

HEALTH.
WATER. SOIL.

Homeseekers! You find these three elements in CHICO VECINO, a subdivision of the famous Rancho Chico, comprising 23,000 acres of fine loaming soil, divided into orchards, vineyards, and grain fields, owned and conducted by Gen. John Bidwell, the pioneer of California.

Health.—Few places outside of CHICO VECINO afford a greater attraction in this respect. It is unqualifiedly a healthy location, as the records will show. Malaria is unknown here, for the reason that irrigation is unnecessary, and stagnant pools, breeding disease germs, are not to be found.

Water.—The northern and southern boundaries of CHICO VECINO, are swiftly flowing, gravelly bottomed streams of clear and pure mountain water.

The banks of these streams are thickly wooded with gigantic oaks, six feet in diameter, the wide-spreading sycamore and the ash, gracefully festooned with the luxuriant wild grape vine, affording the most beautiful drives that human heart could wish.

Soil.—Without exception the best in the State of California, dark and loamy, having an average depth of fifteen feet, with a substratum of gravel through which percolates pure water from the adjoining mountains. These subterranean streams solve the mystery of successful fruit culture in CHICO VECINO without irrigation.

Fruits.—Olives, figs. pomegranates, walnuts, almonds, pecans, grapes, apples, pears, peaches, plums, prunes, apricots, and all kinds of grains and grasses, find here their natural home, and have been successfully cultivated by Gen. Bidwell for the last thirty-five years.

Situation.—CHICO VECINO is situated in Butte County, California, eighty-five miles from Sacramento, the Capital city, and one hundred and eighty-five miles from San Francisco, the great metropolis of the Golden West

The southern boundary of CHICO VECINO separates it from the City of Chico, thus being its neighbor as the name Vecino implies.

The California & Oregon Railroad forms its western boundary. This is the through overland road via the Northern Pacific Railway.

Subdivisions.—CHICO VECINO is subdivided into tracts, from suburban lots 90x200 feet to twenty-acre tracts.

The avenues are eighty and one hundred and fifty feet wide and so arranged that each tract is bounded on all sides by an avenue. Through the centre of this tract is a main Alameda, one hundred and sixty-five feet wide, called the Esplanade, on either side of which are four rows of shade trees, some of which are twenty inches in diameter.

Following the meanderings of Lindo Creek, is another drive one hundred feet wide in the narrowest place, and is being beautified with ornamental shrubbery.

Chico—Is a thriving place of six thousand population and is noted for its fine schools and churches,

A State Normal School, a large three story brick hotel, and other improvements aggregating $250,000, are now in process of construction.

Premium Awards.—Chico holds the Gold Medal for the best citrus exhibit, displayed at Oroville, December, 1887. John Bidwell has been awarded more premiums for best exhibit of fruits, grains and produce at the State, County and Mechanics Fairs than any other individual in the state.

For full particulars, maps, pamphlets etc., address

CAMPER & COSTAR, Agents.

Chico, Butte County, California.

THERMALITO COLONY.

THE PASADENA OF CENTRAL CALIFORNIA.

THERMALITO was surveyed as a Colony site in July, 1887. Its location is adjoining the City of Oroville, Butte County, California, being separated only by the Feather River, a beautiful stream which in the driest season has a flow of not less than 100,000 inches of water. This stream affords abundant opportunities for boating, fishing and bathing, but one mile distant by traveled road from Oroville, a city of 3,000 inhabitants, it affords the dwellers there an opportunity to have a beautiful country seat, a home amid the orange, olive, fig and vine, with all the beauties of a tropic climate.

CLIMATE.

Where the ORANGE grows a temperate climate is assured, for the golden fruit will not flourish or even exist in a lower temperature than 25° above zero. The following table is authentic, having been compiled from Government statistics:

	Average Winter Temperature.	Average Spring Temperature.	Average Summer Temperature.	Average Fall Temperature.	Average Annual Mean Temperature.	North Latitude.
Nice, France	47.8	56	72.3	61.6	59.5	43.45
Florence, Italy	44 3	56.3	74	60	58.9	43.45
Rome, Italy	48 9	57.6	74.2	62	60.7	42
Naples, Italy	48 5	58.5	72.2	64	61.8	40
Valencia, Spain	50 7	63	73.3	66	63.8	39
Palermo, Italy	53	59.3	74.7	68	63.5	38
St. Michael, Azores	57.9	61.2	68.3	62.3	62.3	38
Malta, Sicily	57.5	62.4	78.2	61.6	67.3	36
Algiers	55	66	77	62	65	36
Malaga, Spain	55	68	78	60	65.3	36
Madeira Islands	60 5	62 4	69.6	67 3	65	32.50
Cadiz, Spain	52.9	59 5	70.4	65.4	62.1	36.37
Los Angeles	56.6	57.92	68.51	63.8	61.39	34
Oroville	52.9	64.5	78.8	65.6	65.40	39 27

HEALTH.

The health of Thermalito can not be excelled, lying on a mesa or plateau eighty feet above Feather river; the Colony lands are rolling with a grade towards the river, giving the finest drainage. As a proof of the opinion of the oldest inhabitants on this subject the Board of Supervisors of Butte County selected a site on Thermalito for the County Infirmary, it being the most available situation for health in the County.

PRODUCTIVENESS.

In 1886 the citizens of Oroville determined to enter into the business of Citrus Fruit Culture, and formed a corporation known as the "Oroville Citrus Association," consisting of twenty of the most prominent citizens. After a careful examination of the whole surrounding country, these men selected THERMALITO as the location for their orchard.

This of itself was a verdict in favor of Thermalito, and their faith has proved to be well founded. No more successful venture was ever made. In addition to this you have but to come and see for yourself.

On Thermalito, trees of nine years of age, ladened with the Orange, Lemon, Fig, Apple, Pomegranate, Cherry, Prune, Plum, Olive, Apricot, Peach; in fact, every kind of fruit and vegetable known to a semi-tropic climate are now growing

BEAUTY OF LOCATION.

THERMALITO lies on the south of the far-famed Table Mountains, which rise 1,200 feet above the sea level, and protect it from the cold winds of winter. Feather river on the south and east, the Sacramento valley on the west with the Coast Range Mountains beyond, the Sierra Nevada, Marysville Buttes on the south, and the snow-capped peaks of Mt. Shasta and Lassen 100 miles to the north, all combine to give the Colony the name of "The Beautiful." The eye feasts continually on scenes of majestic grandeur.

WATER SUPPLY.

THERMALITO is possessed of the most complete water supply in the State of California. It controls the entire waters of the west branch of the Feather river. Its supply is 6,600 miners' inches, or a flow of 3,500,000 gallons per hour, a greater supply than the city of San Francisco has. This water is conducted for twenty-five miles through three broad canals to the Colony. The cost of these canals exceeded $300,000 in the beginning. The water is pure and soft. It is supplied to the City of Oroville for all purposes, and is furnished free to purchasers of land in Thermalito for three years after purchase and after that, at the minimum rate so that the cost of irrigation will not exceed one dollar per acre per annum. Thermalito does not depend on an awkward system of ditches for irrigating purposes, but has already laid and has in use nine miles of water main and delivers the water under a pressure of not less than 100 feet on every lot in the Colony so that it can be used for fire as well as household purposes. For quality, quantity and power no place in California can compare with Thermalito's water facilities.

IMPROVEMENTS.

The Company has spent over $100,000 in the last year improving the Colony. The magnificent Bella Vista Hotel will cost when completed $35,000. The water-pipes already in use have cost $25,000. Eleven miles of broad avenues have been graded and are in use. The Grand Avenue for three miles is planted with beautiful ornamental trees, and is the longest pleasure drive ever laid out in Butte County. Purchasers of land have this year planted 300 acres of orange grove and thirty families have located and built lovely homes in Thermalito.

The Company plants and cares for orchards at cost. Cost of an Orange Grove:

10 acres land at $100 per acre,	$1,000
Planting and care for one year, including preparation of ground,	250
690 Seedling Orange Trees, 3 years old, 40 cents,	276
Total,	$1,526

Budded Orange Trees will cost 30 cents per tree more than Seedlings figured above, and if desired 108 trees per acre can be planted by putting the trees 20 feet apart instead of 25 feet as calculated above.

SOIL.

The soil of Thermalito is a rich red clay and gravel, in many places supporting a growth of grand pine and oak trees.

TERMS OF SALE.

Town lots in Thermalito are sold at from $50 to $250 each, being 50 x 150 feet and 90 x 160 feet in size.

Acre property ranges from $50 to $150 per acre, according to distance and location.

WE SELL for one-third cash, one-third in one year, and one-third in two years, with interest on deferred payments at the rate of seven per cent. per annum. FREE WATER for all purposes being given for THREE YEARS to all purchasers

THERMALITO offers the finest opportunity for investment and homes in California. The Great Northern Railroad lines now moving westward must come down the North Fork of the Feather River after passing through Beckwith or Fredonia Pass, and THERMALITO lies at the mouth of the canyon of the North Fork. Any railroad development must cause a rapid rise in values. But this is not necessary, the productiveness of the land will make it pay under cultivation, interest on $2,000 per acre. We cordially invite examination.

THERMALITO COLONY CO.,
OROVILLE, CALIFORNIA.

PORTLAND FROM THE HEIGHTS. MT. HOOD IN THE DISTANCE.

Portland Oregon.

METROPOLIS OF THE PACIFIC NORTHWEST.

Population, • • • • • • 70,000
Altitude (R. R. level), • • • • 58 ft.

THE tourist, after thousands of miles of journeyings amid the magnificent panoramas which mark the trip across the continent, begins to wonder if nature's sketch-book is not exhausted as he approaches the metropolis of the Pacific Northwest from any one of the numerous directions by which it can be reached. He has seen, perhaps, the Garden of the Gods, the peak encircled environs of Manitou, the orange groves of Los Angeles, the blue waters of Puget Sound and the forest-crowned summits of its guardian hills and mountains, caught glimpses, from his Pullman car window, of the unrivaled pastoral beauty of the far-famed valley of the Willamette and comes into Portland seriously debating in his own mind, in all probability, the possibility of finding anything to break the dull monotony of the everlasting round of sight-seeing.

Some years ago, a well-known writer after having paid high tribute to the general beauty of a Northern California county, said, "If the visitor is in search of the sublime let him take the Overland route from Eureka to Ukiah and ask the stage driver to notify him when he reaches the point where Trinity, Mendocino and Humboldt counties 'pool their issues,' and then let him drink his fill of the splendid scene outspread before him and gaze entranced on mountains piled on mountains, rivers running to the sea, and 'vales stretching in pensive quietness between.'"

The tourist may take a much-traveled man's word for it that from Portland Heights he will see all that the writer above quoted so graphically epitomized, with the added charms of all that a virile civilization can do to smooth down the rugged asperities of "nature unadorned." At his very feet a proud young city, mistress of a commerce which makes far-off continents and the isles of the sea tributary to her growth and prosperity, asserts her unquestioned right to the title of sovereign of the occident, so far as the great Northwest empire is concerned, for such is the geographical position of Portland, virtually located at the confluence of the two great rivers of the Northwest, that while rivalry is possible, the attempt to deprive her of supremacy is labor wasted and time thrown away.

The smoke of the factories, the muffled roar of machinery, the masts of sea-

"MULTNOMAH FALLS," COLUMBIA RIVER.

going ships lying at her wharves, the spires of churches, the turrets of public school-houses, streets crowded with trucks, drays, hacks and cabs, busy throngs of business men and elegantly-dressed women, swarthy Italian, ruddy Dane, nervous American, phlegmatic Englishmen, stolid Mongol, mercurial Celt, all sorts and conditions of men, in short, elbowing their various ways in the pursuit of the almighty dollar, make up a cosmopolitan picture of metropolitan life, which, finding it as the tourist does on the outermost rim of the far West, is well worth the study of the most blase globe trotter that ever yawned over his "chops and shandygaff" at Brookes', Delmonico or anywhere else, for that matter. For here is an object lesson of the irresistible march and growth and progress of the great and glorious Yankee nation which is full of suggestion. And that it is suggestive is well witnessed by an incident which came under the observation of the writer quite recently. A very distinguished and prominent citizen of a Southern state, having written to a former fellow-townsman for information regarding Portland as a place for investment, couched his inquiries in such terms as to lead to the inference that he supposed Portland to be merely an outlying frontier settlement. In reply he was told that right here in this Western town he would find business blocks, whole squares of them, superior to anything in New Orleans. And this is true. And, moreover the stocks kept on hand, both in quality and quantity, are in keeping with outside appearances. A writer, in the *Daily Hotel and Commercial Advertiser*, of Portland, gives the reading public this bird's eye view of the goods and wares found for sale in Portland stores: "Shells from the isles of the sea, shawls from farthest India, curious lacquerie from China and Japan, faience from Limoges, cutlery from Sheffield, silks from Lyons, delicatessen from Germany, caviare from the land of the Tsar, olives from Spain, lemons from Sicily, charms and amulets from Palestine, lingerie from Paris, gems from all mines, diamonds, opals, rubies, pearls, emeralds and all others, Swiss marvels of mechanical ingenuity, oranges and pomegranates from California, wine from all nations and bourbon from the blue grass region, everything that Yankee skill invents, makes or trades in, from a pin to a combined reaper and thresher, each and all are to be had for money or approved paper. And to these may be added the apples, pears, plums, melons, wool, grain, game, fish, lumber and iron and gold and silver and the innumerable other products of the marvelously rich commonwealth of which Portland is, and for all time to come will be, the great head centre and metropolis." This is a brief, but it is by no means a comprehensive, and in no sense of the word an exhaustive resumé of the metropolitan aspects Portland presents in its minor commercial relations. For, in view of the fact that for the most part the articles above enumerated relate almost exclusively to the retail trade of the city, they may well be called the minor details. The export trade of Portland amounts to over fifteen millions of dollars annually, the principal articles being grain, wool, fish and lumber, her merchants having correspondents in the British empire, Peru, Chili, France, China and Amman, with Australia and Japan soon to be added to the list. With respect to manufactures a recent tabulation of the industries of the city and outlying suburbs by Mr. L. H. Wells, editor of the *West Shore*, shows an invested capital of $10,457,000, an employed force of 4,891 laborers, a wage roll of $2,693,573, and an annual product valued at over $17,000,000. Such is Portland in some of its merely material aspects. And its interests always have been, now are and are likely always to be, in the hands of men with a keen eye to the main chance, and fully able to maintain and keep intact the position already won. The writer not long ago fell into conversation with a gentleman who has been in his time a member of the Boards of Trade of Louisville, Ky., and of New Orleans. "Why, sir," remarked the gentleman to the writer, "Your Portland Board of Trade are a smart—a remarkably smart—body of men. I have been accustomed to disputes,

CASTLE ROCK, COLUMBIA RIVER.

THE TOOTH BRIDGE, COLUMBIA RIVER.

to long and excited debates, from which, very frequently, nothing practical resulted. Here the members of the Board meet, and seem to know just exactly what is to be done and how to do it. A member gets up and coolly, dispassionately presents his facts and figures. Others do the same, a conclusion is arrived at without jar or jangle. All pull together as near as may be. It is Portland first and the rest of the world afterward, and it is no wonder the city is forging ahead. It has doubled its population in less than ten years, and I see no reason why the rate of increase should not be more rapid in the future than in the past."

In view of these facts Portland, therefore, presents an instructive and interesting object lesson to the student of urban growth and of the influence of American ideas and American institutions upon progress and development.

But the tourist, however much of a Gradgrind he may be in the matter of facts and figures, wants something more than mere numerals to make his visit enjoyable. While on Portland Heights, if his eyes have not been idle, he has seen far away to the east Mount Hood, the ancient, white-robed sentinel of the Columbia, lifting its regal splendors far aloft, while still more distant, to the north and south, other snow-clad peaks, scarcely inferior in grandeur, keep watch and ward over forest and field, lake and river, city, town, hamlet and solitary farm house. This scene once witnessed will never be forgotten. Descending from the Heights, the homes of Portland, tree embowered, lawn surrounded, from the modern palace of the millionaire to the ornate cottage of the thrifty citizen, may well challenge inspection, and, not that alone, but critical comparison, with those of any city of thrice its size and much closer proximity to the great centres of wealth and population. Whether from the East, or North, or South, the tourist will see in park and by the roadside familiar forest growths which give a homelike aspect to all his surroundings. All of the cherished companions of the flower gardens of the temperate zone greet him at almost every step, from door yard or wide and perennially verdant lawns. Portland's magnificent High School building, by far the finest and most ornate on the Pacific slope, forms one of the most attractive features of the city. Near by, the immense structure to be devoted to a permanent industrial exhibition of the arts and industries of the Pacific will furnish additional evidence, if needed, of the progressive tendencies of the metropolis of the Northwest. The handsome, castellated Armory of the citizen soldiery of the metropolis is well worth a visit. In the art stores of the city admirable reproductions of the splendid scenery of the Northwest by local artists furnish ample reasons for many a visit, many an hour of pleasant study, and the liberal expenditure of money for souvenirs of the tourist's sojourn. The game and fish and fruit and vegetable stalls of the city speak more intelligibly and instructively of the fecundity of the soil and waters of the Northwest than a score of volumes of descriptive writing could do, and give most appetizing hints to the man or woman fond of the good things of this life of the wide range of choice given to the caterer for supplies for his table. Near the post-office the tourist will see, the pride of the metropolis, the "Hotel Portland," a magnificent building, which, when completed and furnished, will cost between $800,000 and $1,000,000. This splendid caravansary, to be complete in all its appointments, with all modern conveniences from corner to cope stone, will be ready for the reception and entertainment of tourists by December, 1889, and once housed in its comfortable apartments the visitor will doubtless be loth to seek other and possibly inferior accommodations. Not only Portland Heights and the parks and private grounds and public buildings of Portland invite the leisurely inspection of the tourist, but along the city front for miles there are abundant opportunities to pleasantly while away the leisure hours. A score or two of trains arrive and depart daily. The river is hourly vexed by the arrival and departure of a fleet of steamers. In the shipping season the long wharves are

COFFIN ROCK, INDIAN BURIAL PLACE, LOWER COLUMBIA RIVER.

lined with deep sea ships and immense warehouses groan with the produce of an agricultural, fruit-growing, wool-producing and mineral empire, as yet but in the infancy of its development. When the day's sight-seeing is over, well-equipped opera houses will be open to the lovers of the music and of the drama.

If a day's or a week's outing is desired, while it can not be truthfully said that "all the world" is before the tourist, "where to choose," there is within easy reach of Portland by well-equipped steamer, or by rail, an almost limitless variety of scenic and restful attractions. An hour or two's ride by steamer down the Willamette and up the Columbia, or half that time's ride by motor road, will take the visitor to the Garrison town of Vancouver, W. T., where the parade of the regiment on duty and the afternoon's music by the band, will round out a day of rare enjoyment. Another day, or more, it may be, devoted to a trip up the Columbia, the scenic splendors along which dwarf almost into littleness those along the Hudson and the Rhine. Multnomah Falls, pyramidal rocks to which Egyptian obelisks are mere toys by comparison, Cape Horn, a natural fortress of the Titans of the mountains, the Cascades, the Pillars of Hercules, the broad majestic river, the castellated, forest crowned hills, and a score of cataracts leaping from dizzy heights are but incidents in the ever-varying charms of this memorable trip.

Within an hour's ride of the city by steamer or rail the Falls of the Willamette, which not even the busy manufacturing town of Oregon City utilizing their unrivaled water power can render prosaic, can not fail to attract the tourist and well repay the time and trouble of a visit. All around the city are cosy nooks in the recesses of the darkling hills or on the banks of the Willamette and its tributary streams are retreats where the picnicker, with well-filled hamper, to which with little trouble can be added fish from the stream and game from the woods, can take his ease in the shade of oak and pine and maple, and dream, if he or she likes, of the forest of Ardennes and realize to the full that their lives, for the nonce at least, "remote from public haunts," are pleasanter than amid the rush and roar and rattle of commerce and trade and politics and manufactures. Within a few hours' travel by rail or steamer in three or four directions, "the salt air of the sea," which almost as much as sleep is "tired nature's sweet restorer," "balm of hurt minds" and general cure all for half the ills that flesh is heir to, is delightfully accessible and all along a hundred or two hundred miles of sea coast good hotel accommodations and unrivaled facilities for camping out and "roughing it" are to be had at reasonable rates. Take it all together, it may truthfully be asserted that in point of scenic attractions, delightful climate for the greater part of the year, reasonable and bountiful facilities for enjoying the true "*dolce far niente*" of the tourist Portland offers unsurpassed attractions. It is by no means to be forgotten that club life has its votaries in Portland as elsewhere. Besides two or three social organizations on the regular lines, the city boasts of an association known already far and wide as the Alpine Club, devoted to the material and scientific development of the state. Its membership already comprises much of the solid wealth and intellectual force of the state and a rapidly accumulating cabinet of the mineralogical and historical curios of the Northwest, makes it certain that the open sesame to its hospitable doors will give the tourist an unrivaled opportunity to acquaint himself at slight trouble with much of interest concerning the Northwest, which otherwise he could obtain the knowledge of only at great trouble and expense.

The tourist, then, is cordially invited to Portland with the confident assurance that the longer his stay and the more thorough his exploration of its surroundings may be, the more he will be charmed with the locality and the more likely he will be to comprehend the pride which its denizens feel in, and the affection they bear for, the Metropolis of the Pacific Northwest.

CONTENTS.

	PAGE.
From the Missouri River to Denver,	9
From Denver to Pueblo,	13
Pueblo to Ogden,	31
Pueblo to Alamosa,	69
Alamosa to Española and Santa Fé,	81
Alamosa to Silverton,	89
Silverton to Montrose,	103
Salida to Aspen,	109
Leadville to Dillon,	123
Ogden to San Francisco,	127
San Francisco to San Diego,	149
Los Angeles to Santa Barbara,	171
Los Angeles to San Diego,	178
Saunterings Around San Francisco,	201
To the Yosemite,	215
From San Francisco to the Great Northwest,	219
Index to Stations on Denver & Rio Grande and Denver & Rio Grande Western R. R.,	250
Mountain Peaks of Colorado,	252
Mountain Passes of Colorado,	252
Elevation of Lakes,	252
Altitudes of Towns and Cities,	253
Distances from Denver,	253
Pronunciation of Proper Names,	253
Index to Stations on the Southern Pacific R. R.,	254
Addenda,	260

ILLUSTRATIONS.

	PAGE.
Arapahoe County Court House, Denver,	15
Alignment of the Denver & Rio Grande R.R. over Marshall Pass,	38
Alignment of Toltec Gorge District,	91
A Rocky Mountain Beauty Spot,	123
Approach to the Black Cañon,	46
A Donkey Brigade,	30
Along the Animas River,	98
Animas Cañon and the Needle Mountain,	102
A Quiet Nook,	57
A Ute Council Fire,	51
A Typical Mexican,	83
Assembly Hall, Tabernacle and Temple, Salt Lake City,	65
Approaches to Oakland Ferry,	148
At the Golden Gate,	134
Artesian Well, South Riverside,	294
Bird's-Eye View of Denver,	6
Bird's Eye View of Aspen,	120
Bird's-Eye View of Salida,	108
Bird's-Eye View of Ouray,	106
Brown's Cañon,	109
Bird's-Eye View of Salt Lake City,	61
Bee Hive House,	66
Bird's-Eye View of San Francisco,	152
Big Trees of Calaveras,	218
Bath House, Cicero Place, Green Lake,	328
Colorado's Capital Building, Denver,	16
Currecanti Needle, Black Cañon,	48
Crested Butte Mountain and Lake,	41
Cañon of Rio de Las Animas,	97
Cascades of the Blue,	126
Castle of the Cliff Dwellers, Mancos Cañon,	94
Climbing the Mountains at Veta Pass,	72
Chippewa Falls in the Black Cañon,	50
Chiefs of the Uncompahgre Utes,	107
Castle Gate,	54
Capitol Building, Sacramento,	144
California's Mammoth Grape Vine,	200
Concrete Pipe Works, Pomona,	264
Cascade in San Antonio Cañon, Ontario,	192
Cape Horn, Columbia River,	234
Castle Rock, Columbia River,	310
Coffin Rock, Indian Burial Place, Lower Columbia River,	312
Cicero Place, on Green Lake, Near Seattle,	242
Depot at Colorado Springs,	20
Donner and Webber Lakes,	140
Dip-Net Fishing at the Dalles of the Columbia,	238

ILLUSTRATIONS.

	PAGE.
Embudo, Rio Grande Valley,	81
Exploring the Walls,	116
Euclid Avenue, Ontario,	190, 286
Echo Rock,	101
Fremont Pass,	125
First Congregational Church, Denver,	14
Falls of the Yosemite,	216
From Nimbus Knob to the Sacramento,	302
Forests of the Columbia,	236
Glimpses of Pike's Peak,	26
Gateway to the Garden of the Gods.	22
Grape Creek Cañon,	32
Grand Cañon of the Colorado,	53
Grand Cañon, from To-Ro-Wasp,	68
Gate of Ladore,	44
Garfield Memorial,	93
"Genl. Fremont" Big Tree, and Beach View,	207
Glimpse of Celestial Life in San Francisco,	156
High School Building, Denver,	10
Hotel and Grounds, Ontario,	282
Hotel Temescal, South Riverside,	292
In the Garden of the Gods,	27
In San Francisco Bay,	154
In the Semi-Tropic Zone, Los Angeles,	170
Inside and Outside Headers—Del Monte,	212
Jenner Falls,	247
Lick Observatory, Mt. Hamilton,	205
Lower Cape Horn, Columbia River,	234
Manitou Springs and Pike's Peak,	24
Mount of the Holy Cross,	114
Marshall Pass—Eastern Slope,	35
Marshall Pass—Western Slope,	42
Marble Cañon,	122
Main Street, Ontario,	280
Mariposa Big Tree Grove,	220
Mount Shasta,	228
Mt. Rainier,	249
"Multnomah Falls," Columbia River,	308
Megdenhour Bay and Edgewater Point, near Seattle,	244
New Mexican Life,	86
Near San Gabriel and Pasadena,	179
Napa Soda Springs,	223
Over the Range,	Frontispiece.
On the Lookout,	96
Old Church of San Juan,	82
On the Uncompahgre,	104
On Wheels, Through Golden Gate Park,	160
On the Beach at Santa Cruz,	266
On the Rio Chico,	223
Orange Orchard, Pomona,	196
Ocean Sculpture, Santa Monica.	198

ILLUSTRATIONS.

	PAGE.
Old Mission Church at Santa Barbara,	174
Old Mission Church, Monterey,	202
Over the Range,	130
On the Balcony, Hotel del Monte,	210
Palmer Lake,	18
Phantom Curve,	91
Pueblo de Taos,	85
Pueblo Indians,	84
Pavilion, Woodward's Gardens,	158
People's Bank, Pomona,	266
Portland from the Heights. Mt. Hood in the Distance,	306
Portland from East Bank of the Willamette,	233
Pacific Avenue, Tacoma, 1877,	240
Pacific Avenue, Tacoma, 1888,	240
Queen's Cañon,	67
Rainbow Falls,	21
Rooster Rock, Columbia River,	237
Sierra Blanca,	74
Sangre de Cristo Range, from Marshall Pass,	40
Silverton and the Sultan Mountain,	100
Summit of Veta Mountain,	76
Spanish Fork Cañon,	58
San Antonio Falls, Ontario,	193
State Bank Block, Ontario,	283
Scene in San Antonio Cañon, Ontario,	186
Sierra Madre Villa,	271
Sir Joseph Hooker Oak, Chico Vecino,	221
The Royal Gorge,	33, 36
The Seven Falls, Cheyenne Cañon,	25
Trout Fishing at Wagon Wheel Gap,	79
Trout Fishing on the Cimarron,	49
Toltec Gorge and Tunnel,	92
Tramway in Little Cottonwood Cañon,	59
The Great Salt Lake,	63
Tahoe Scenery,	138
The Loop,	166
The Petrified Forest,	163
Twin Falls,	225
"The Old Cabin on the Columbia,"	230
Tooth Bridge, Columbia River,	310
Upper Twin Lakes,	110
Up the Rio Grande,	78
Union Block, Seattle,	326
View of Fourteenth Street, Denver,	8
View of Public Buildings, Denver,	12
Veta Pass and Dump Mountain,	70
Views from the Cliff House,	151
View in San Antonio Cañon, Ontario,	184
Wagon Wheel Gap,	77
Yosemite Valley,	214
Young America's Friend,	189

INDEX

OF TOWNS AND POINTS OF INTEREST.

	PAGE.		PAGE.
Acequia	17	Burnham	16
Alta	60	Byron	161
Alta Branch	60	Byron Hot Springs	151
Alamosa	75	By Rail to Los Angeles	175
Alcazar The, S. F.	155	Castle Gate	54
Alcatraz Island and Angel Island.	157	California State Mining Bureau, S.F.	157
Alhambra	197	Cameron	167
Alviso	204	Caliente	195
Albany	281	Carpinteria	175
Alessandro	296-297	Camulos	177
American Fork	59	Calumet Branch	111
Amargo	93	Cañon of the Grand River	116
American River Bridge	145	Carbondale	119
Antelope Springs	80	Carlin	132
Antonito	82	Carson Lake	136
Antioch	161	Cascade	142
Animas Cañon	99	Cape Horn	143
Anderson	225	Castroville	209
Army Point	147	Calaveras Grove, The	217-218
Arkansas Valley	31	Castle Rock	17
Aspen	121	Carlisle Springs	29
Ashland	229	Cañon City	31-32
Athlone	162	Cedar Divide	49
Auburn	143	Cedar Pass	131
Avalanche Creek	119	Chippeta Fall	47
Aztec	96	Chama	93
Azusa	182	Chinese Theatre, S. F.	155
Barranca	83	Chattanooga	103
Baldwin Theatre, S. F.	165	Chico	222
Banks, S. F.	157	Chico Vecino	223, 303
Banta	161	Cheyenne Mountain	21
Bakersfield	165	Cimarron	47
Bathing.Pool, The	118	Cimarron Cañon	49
Bathing, Accommodations For	118	Cliff Dwellings	95
Battle Mountain	133	Cliff House	150, 157
Benicia	147	Claremont	182
Bentwood	161	Climate, The	191
Bethany	161	Cluro	133
Berenda	162	Climate of Puget Sound, The	241
Bear Creek Falls	104	Clark's Magnetic Spring	29
Beowawe	133	Coal Basin	119
Bingham	60	Corinne	128
Bingham Junction	60	Colfax	143
Bingham Branch	60	Colorado Springs	21
Bijou Theatre, S. F.	155	Colorado City	23
Big Trees, The	218	Coal Creek	31
Black Cañon of the Gunnison	47	Coal Creek Branch	31
Bloomfield	96	Colorado Desert, The	52
Book Cliffs, The	52	Coal Branch	55
Brown's Cañon	109	Comanche Cañon	84
Brigham	128	Coal Mines	161
Browns	135	Compton	171
Bush Street Theatre, S. F.	155	Colton	183-287-289
Buena Vista	111	Coronado	189

INDEX.

	PAGE.
Cocamonga	191
Cottonwood Springs	111
Crane's Park	115
Crested Butte	43
Crested Butte Branch	43
Crystal	121
Currecanti Needle	47
Cuchara Junction	76
Cumbres	93
Davis	146
Dallas	107
Denver	13-15
Delta	51
Del Norte	76
Descending to the Desert	167
Desert	136
Del Monte	209
Dillon	126
Divide	231
Doubling on our track	159
Donner Lake	141
Douglass	17
Durango	95
Duarte	182
Dutch Flat	142
Eastern Railway Lines, S. F.	157
East Riverside	183
Eagle River Cañon	115
El Moro	69
Elmira	147
Elk Park	104
Elk Mountain Railway	119
Elko	132
Embudo	86
Emigrant Gap	142
Espanola	86
Espanola to Santa Fe	87
Etiwanda	182
Eugene	231
Express Office, S. F.	157
Extension of the D. & R. G. R. R.	119
Farmington	96
Fair Oaks	201
Fertile Valleys	176
Fertile Valley, A	183
Florence	31, 171
Fort Dushane	54
Fort Lewis	97
Fountain	27
Fruitvale	52
Fresno	162
Fremont Pass	124
Garfield Beach	67
Garland	73
Garfield Memorial	91
Geological Features	127
Glendora	182
Glenwood Springs	117-118
Glen Park	17
Golden Gate Park, S. F.	150-155
Goshen Division, The	164
Goshen	164
Golconda	133

	PAGE.
Grape Creek Cañon	34
Grand Junction	52
Grand Cañon of the Colorado	54
Great Salt Lake	62
Granite	111
Gravelly Ford	133
Green River	53
Gunnison	43
Hack Fare, S. F.	157
Harbor, A Magnificent	243
Historic Ground	142
Hot Sping	32, 62, 136
Hotel del Monte, The	211
Humboldt	135
Humboldt Lake	136
Ignacio	95
Independence Lake	141
Irrigation in the Artesian Belt	165
Ironton	103
Jacksonville	229
Jordon River	60
Junction	145
King's River	164
Kingsburg	164
Keton	129
Kyune	55
Lake City	45
Lake City Branch	45
Lake Fork Cañon	45
Lake Park	67
La Veta	71
La Jara	81
Lathrop	161
Lamanda Park	180, 269, 270
Lake El inore	185
Lake Region, The	136
Lake Tahoe	138, 139
Lake Colony	298-299
Lehigh	59
Leland Stanford Jr., University	204
Leadville	112
Lick Observatory, The	207
Lincoln	219
Littleton	17
Lower Crossing	54
Los Piños Valley	89
Los Angeles	169-168
Los Angeles to Santa Barbara	171
Los Angeles to San Diego	178
Lordsburg	182
Los Angeles, Returning to	191
Long Beach	201, 277
Los Gatos	207
Marshall Pass	41
Marshall Pass Station	41
Manasa	82
Martinez	159
Madera	162
Markets, S. F.	157
Magnificent Scenery	99
Malta	112
Marysville	219
Manufacturing	236

350

INDEX.

	PAGE.
Manitou	23-25
Mears Junction	39
Merchants Exchange, S. F.	157
Merced	162
Menlo Park	203
Mission Dolores, The, S. F.	150-157
Millbrae	201
Mirage	135
Military Post	16
Monarch Branch	37
Montrose	52
Monte Vista	75
Modesto	162
Mojave Desert, The	167
Mojave	167
Montalvo	176
Monrovia	272, 182
Mount of the Holy Cross	124, 113
Monument	129
Moors	131
Monterey	213, 300-301
Modern Improvements	235
Monument	19
Monument Park	19
Murietta	185
Mud Lake	126
Muir's Peak	227
National City	188
Newhall	168
Nevada Desert, The	135
Newcastle	143
North Cocamonga	182
North Ontario	182
North Pomona	182
Oakland Pier	147
Oakland	159
Ocean Side	185
Ogden	127
Ojo Caliente	83
Ojai Valley, The	176
Old Mission, The	175
Ontario	191-194, 279, 284
"Orpheum" Opera House, S. F.	155
Ortega	175
Oreana	135
Oroville	221
Oregon City	232
Ouray	105
Ouray to Montrose	107
Overland Park	16
Palmer Lake	17
Parnassus Springs	29
Palm Valley	291
Parkdale	37
Palmilla	82
Pagosa Springs	93
Pacific Slope, The	95
Panorama Building, S. F.	155
Pasadena	178, 261-263
Palisades of the Humboldt	132
Palisade	132
Pacific Grove	213
Penny's Hot Springs	119
Perry Park	17
Phantom Curve	89
Pinole	147
Picturesque Surroundings	236
Pike's Peak	25-27
Pleasant Valley Junction	55
Placer	72
Places of Interest near Española	86
Plains Region, The	145
Plains Across, The	9
Petersburg	16
Poncha	39
Poncha Pass	39
Poncha Springs	39
Port Costa	147
Post office, S. F.	157
Point Fermin	172
Pomona	194, 265-267
Portland	232-235, 307-313
Price	54
Provo	58
Presidio Reservation, S. F.	157
Prospect	119
Promontory	128
Pueblo of San Juan	86
Pueblo de Taos	87
Pueblo of Santa Clara	87
Pueblo	27-29
Pyramid Lake	136
Raymond	178
Red Narrows	56
Redwood	203
Red Cliff Cañon	115
Red Cliff	115
Reno	137
Red Bluff	223
Redding	226
Rialto	182
Riverside	185
Riverside	111
Royal Gorge	34
Roseville	189
Robinson's Lake	121
Rocklin	143
Rogue River Valley	229
Roseburg	231
Rocky Mountains, The	11
Rye Patch	135
Salida	37
San Luis Branch	39
Sapinero	45
Salt Lake City	60
Salt Lake to Ogden	67
San Luis Park	73
Santa Cruz	86
Santa Fé	87
Sacramento	145
San Francisco	149
San Francisco Bay	151
San Joaquin Valley, The	161
San Fernando Tunnel	168
San Fernando	168
San Pedro	171

INDEX

	PAGE		PAGE
Santa Barbara	173, 172	Tehachapi Summit	167
San Pedro to Santa Barbara	172	Tennessee Pass	113
San Clemeth Island	172	Tehama	223
Santa Catalina Island	172	Terminal and Shipping Facilities	243
San Buenaventura	175	Tea Trade with the Orient	245
Santa Paula	177	Thermalito Colony	304-305
Saticoy	175	Tivoli Opera House, S. F.	155
Saugus	178	Tipton	165
San Gabriel Valley, The	181	Tomichi Meadows	43
San Bernardino	182	Toltec Gorge	91
San Dimas	182	Towns in the Desert	131
San Diego	187	Toano	131
San Diego Bay	188	Trinidad	69
San Gabriel	197	Trinidad Branch	69
Santa Monica	197 199, 274	Trout Fishing in the Rio Grande	80
San Mateo	201	Tracy	161
San Bruno	201	Traver	164
San Francisco to Monterey	201	Trimble Hot Springs	97
Santa Clara	204	Truckee	137
San Jose	204-205	Trade with S. America and Mexico	241
Santa Cruz	209, 207	Trade with the Middle West	243
Salem	231	Tulare County, Resources of	164
Scenic Attraction	231, 121	Tulare	165
Seal Rocks	150-157	Twin Lakes	111
Selma	164	Umpqua, Valley of the	231
Seattle	245-249	United States Mint, S. F.	157
Seattle, Advantages of	246	Utah Valley	56
Seattle, Beauty of the City	248	Utah Lake	59
Sedalia	17	Vallejo Junction	147
Shasta	227	Valley of the Eagle	115
Silver Cliff Branch	34	Vapor Caves, The	119
Sierra Blanca	73	Valley of the Humboldt	132
Sierra Madre Villa	181	Valley Region, The	133
Silverton	99	Veta Pass	71
Sierra Nevada Range, The	141, 137	Villa Grove	39
Sisson	227	Visalia	164
Siskiyou Station	229	Waunita Hot Springs, The	43
Soldier Summit	56	Wasatch Range The	54
Southward Bound	159	Walsens	71
South Pasadena	178	Wagon Wheel Gap Branch	75
South Riverside	183, 293 295	Wagon Wheel Gap	78
Source of the Arkansas	124	Wadsworth	136
Soda Springs	142 227	Walker's Lake	136
Spanish Fork	56	West Cliff	34
Spanish Fork Cañon	56	Wellsville Hot Springs	37
Springville	56	Webster	146
Spanish Peaks	69	Wells	131
Straits of Carquinez, Crossing the	147	Webber Lake	141
Stock Exchange, S. F.	157	Wheatland	219
Street Car Fares, S. F.	157	Wilson's College	171
Stage Ride, a Romantic	104	Wildomar	185
State Line, The	229	Winnemucca	133
Sutro Heights, S. F.	157	Winnemucca Lake	136
Summit	142	Wild Scenery	226
Sweetwater Dam, The	191	Woodward's Gardens, S. F.	153-155
Tacoma	237, 243	Yosemite	215
Tehachapi Pass	167	Yuba River, The	219

20 GEMSTONES

CUT AND POLISHED FOR JEWELRY MOUNTING,

STANLEY WOOD,
EDITOR OF
THE
GREAT DIVIDE

The Successful
Monthly
of the
Wild
and
Woolly
West.

ALL FREE WITH THE GREAT DIVIDE.

THE GREAT DIVIDE GIVES SOMETHING FOR NOTHING.

Viz: Cameo, Goldstone, Tiger Eye, Ribbon Agate, Green and Fancy Crocidolites, Carnelian, Jewel Agate, Mosaic, Satin Spar (the peer of Moonstone), Montana Moss Agate, Agate for sleeve buttons, Green Moss Agate, Striped Agate for ladies' brooch, Petrified Wood, Etc., given free as a premium to each new subscriber, if **$1.00**, price of yearly subscription is sent within **30** days of the date of this journal. Each Gemstone is honestly worth **50** cents, and some cannot be bought for **$1** each of any jeweler, and the total value is over **$10**. You naturally say, "Can this be true?" We positively guarantee to refund your money if you are not fully satisfied.

Our reason for offering this costly premium is: *We must advertise to get others to advertise with us, and by this method we will have a national circulation quicker than by any other way that we know of, and our conclusions are sustained by experiments.* **THE GREAT DIVIDE** is a monthly journal, illustrated and printed in an elegant manner; each number will contain articles on Rocky Mountain scenery, minerals, mines, crystals, cliff-dwellers, Indians and their customs, haunts of fish and game, natural wonders, caves, grotesque and marvelous works of nature, burning rocks, mineral springs, climate, wild flowers, and hosts of other interesting things. Brimful of fresh, original and spicy reading every month. Different from any other publication in the world. In addition to the above, Art Supplements in seven colors, which are truly worth training, are frequent features of this original journal.

OUR CONTRIBUTORS are Literateurs, Plain People, Cowboys, Scouts, Miners, Indians; in other words, people familiar whereof they write, and who tell their stories in their own quaint way. You cannot afford to miss this. Sample copy only **10 cents.**

MARVELOUS as this inducement seems, you may rest assured it is genuine or the publishers of this book would not print this advertisement, therefore send **$1.00** to-day for a year's subscription, and the 20 Gemstones will be sent the same day your order is received. Address,

THE GREAT DIVIDE, 1516-1518 Arapahoe St., Denver, Colo.

www.ingramcontent.com/pod-product-compliance
Lightning Source LLC
Chambersburg PA
CBHW031906220426
43663CB00006B/789